Physical Methods in Chemistry and Nano Science. Volume 2: Physical and Thermal Methods

Physical Methods in Chemistry and Nano Science.
Volume 2: Physical and Thermal Methods

Editor

Andrew R. Barron

Contributors

Wala Algozeeb, Andrew R. Barron, Kyle Bayliff,
Christopher Burbridge, Gaowei Chen, Stuart J. Corr,
Caoimhe de Fréin, Johnathan Dietz, Vanessa Espinoza,
Chih-Chau Hwang, Nina Hwang, Anjli Kumar, Xianyu Li,
Yilun Li, Yen-Hao Lin, Samuel Maguire-Boyle,
Shawdon Molavi, Sehmus Ozden, Tawana Robinson,
Pavan M. V. Raja, Desmond Schipper, Basil Shadfan,
Nikolaos Soultanidis, Juan Velazquez,
Kendahl L Walz Mitra, Julia Zhao

MiDAS Green Innovations
2020

First Printing: 2020

ISBN 978-1-8380085-5-0

MiDAS Green Innovation, Ltd
Swansea, SA1 8RD, UK

www.midasgreeninnovation.com

Dedication

To all my postdoctoral researchers, thank you for making me look good.

"Did you ever feel that just by jumping on a plane, all your cares would drop away"

Marillion (1991)

Contents

Acknowledgements

I would like to thank all the contributors to this Volume, for their interest in creating a user-friendly text for their peers.

The myriad collaborators of whom I have had the pleasure of learning from over the years are to be thanked for bringing new characterization methods to my research - you know who you are.

Last, but not least, I would like to thank my wife, Merrie, for putting up with me during the Editing of this book while 'social distancing' during the COVID-19 pandemic.

Preface

This Series intended as a survey of research techniques used in modern chemistry, materials science, and nanoscience. The topics are grouped into volumes, not be method *per se*, but with regard to the type of information that can be obtained. Thus, the Volumes are ordered as follows:

- Elemental composition.
- Physical and thermal analysis.
- Chromatography
- Chemical speciation.
- Molecular and solid state structure.
- Surface morphology and structure at the nanoscale.
- Device performance.
- Applications of analytical methods

As a consequence of this organization methods can be found in different Volumes. For example, X-ray photoelectron spectroscopy is included under Elemental Composition (Volume 1) with regard to its use for determining the chemical composition, while it is included under Chemical Speciation (Volume 3) with regard to determining the identity of component chemical moieties.

The goal was to create simple to understand explanations of methods that allow the reader to gain the knowledge to correctly apply a technique or interpret data. As a consequence, the topics in this book have been developed in partnership with undergraduate and postgraduate students at Rice University over a 7-year period, and because of this there is some variation in depth and focus given to each topic. I make no apology for this diversity.

Chapter 1: Melting Point Analysis

Shawdon Molavi and Andrew R. Barron

Introduction

Melting point (Mp) is a quick and easy analysis that may be used to qualitatively identify relatively pure samples (approximately <10% impurities). It is also possible to use this analysis to quantitatively determine purity. Melting point analysis, as the name suggests, characterizes the melting point, a stable physical property, of a sample in a straightforward manner, which can then be used to identify the sample.

Equipment

Although different designs of apparatus exist, they all have some sort of heating or heat transfer medium with a control, a thermometer, and often a backlight and magnifying lens to assist in observing melting (Figure 1.1). Most models today utilize capillary tubes containing the sample submerged in a heated oil bath. The sample is viewed with a simple magnifying lens. Some new models have digital thermometers and controls and even allow for programming. Programming allows more precise control over the starting temperature, ending temperature and the rate of change of the temperature.

Sample preparation

For melting point analysis, preparation is straight forward. The sample must be thoroughly dried and relatively pure (<10% impurities). The dry sample should then be packed into a melting point analysis capillary tube, which is simply a glass capillary tube with only one open end. Only 1 to 3 mm of sample is needed for sufficient analysis. The sample needs to be packed down into the closed end of the tube. This may be done by gently tapping the tube or dropping it upright onto a hard surface (Figure 1.2). Some apparatuses have a vibrator to assist in packing the sample. Finally, the tube should be placed into the machine. Some models can accommodate multiple samples.

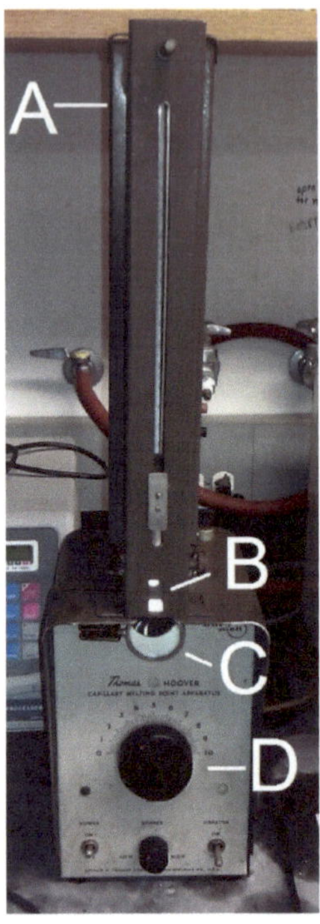

Figure 1.1: A Thomas Hoover melting point apparatus. The tower (a) contains a thermometer with a reflective view (b), so that the sample and temperature may be monitored simultaneously. The magnifying lens (c) allows better viewing of samples and lies above the heat controller (d).

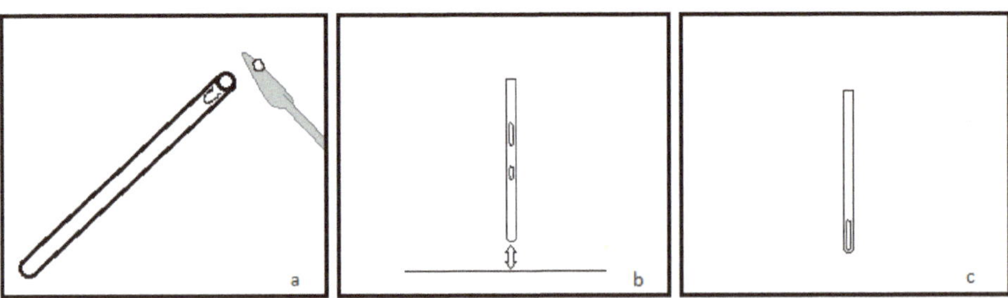

Figure 1.2: Schematic showing how to pack dried sample into a melting point analysis capillary tube: (a) using a spatula, push a sufficient amount of sample into the tube opening, (b) using a tapping motion or dropping the tube, pack the sample into the closed end, (c) the sample is ready to be loaded into the apparatus.

Recording data

Performing analysis is different from machine to machine, but the overall process is the same. If possible, choose a starting temperature, ending temperature, and rate of change of temperature. If the identity of the sample is known, base the starting and ending temperatures from the known melting point of the chemical, providing margins on both sides of the range. If using a model without programming, simply turn on the machine and monitor the rate of temperature change manually.

Visually inspect the sample as it heats. Once melting begins, note the temperature. When the sample is completely melted, note the temperature again. That is the melting point range for the sample. Pure samples typically have a $1 - 2\ °C$ melting point range, however, this may be broadened due to colligative properties.

Interpreting data

There are two primary uses of melting point analysis data. The first is for qualitative identification of the sample, and the second is for quantitative purity characterization of the sample.

For identification, compare the experimental melting point range of the unknown to literature values. There are several vast databases of these values. Obtain a pure sample of the suspected chemical and mix a small amount of the unknown with it and conduct melting point analysis again. If a sharp melting point range is observed at similar temperatures to the literature values, then the unknown has likely been identified correctly. Conversely, if the melting point range is depressed or broadened, which would be due to colligative properties, then the unknown was not successfully identified.

To characterize purity, first the identity of the solvent (the main constituent of the sample) and the identity of the primary solute need to be known. This may be done using other forms of analysis, such as gas chromatography-mass spectroscopy coupled with a database. Because melting point depression is unique between chemicals, a mixed melting curve comparing molar fractions of the two constituents with melting point needs to either be obtained or prepared (Figure 1.3). Simply prepare standards with known molar fraction ratios, then perform melting point analysis on each standard and plot the results. Compare the melting point range of the experimental sample to the curve to identify the approximate molar fractions of the constituents. This sort of purity characterization cannot be performed if there are more than two primary components to the sample.

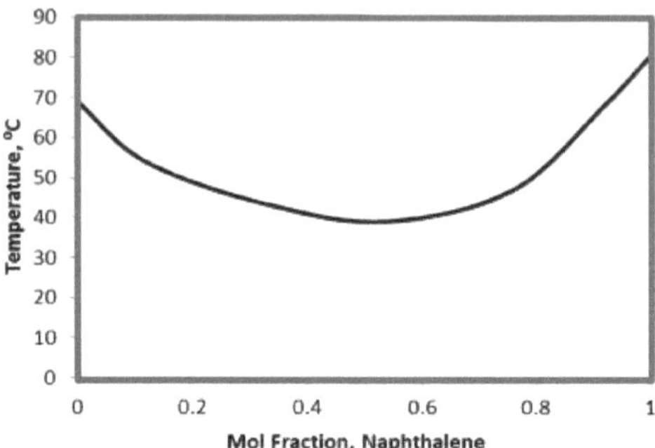

Figure 1.3: A mixed melting curve for naphthalene and biphenyl. Non-pure samples exhibit melting point depression due to colligative properties. Adapted from *Melting Point Analysis*, Chem 211L, Clark College protocol.

Specificity and accuracy

Melting point analysis is fairly specific and accurate given its simplicity. Because melting point is a unique physical characteristic of a substance, melting point analysis does have high specificity. Although, many substances have similar melting points, so having an idea of possible chemicals in mind can greatly narrow down the choices. The thermometers used are also accurate. However, melting point is dependent on pressure as well, so experimental results can vary from literature values, especially at extreme locations, i.e., places of high altitude. The biggest source of error stems from the visual detection of melting by the experimenter. Controlling the change rate and running multiple trials can lessen the degree of error introduced at this step.

Advantages of melting point analysis

Melting point analysis is a quick, relatively easy, and inexpensive preliminary analysis if the sample is already mostly pure and has a suspected identity. Additionally, analysis requires small samples only.

Limitations of melting point analysis

As with any analysis, there are certain drawbacks to melting point analysis. If the sample is not solid, melting point analysis cannot be done. Also, analysis is destructive of the sample. For qualitative identification analysis, there are now more specific and accurate analyses that exist, although they are typically much more expensive. Also, samples with more than one solute cannot be analyzed quantitatively for purity.

Bibliography

IUPUIORGANICCHEM, Exp 3 Melting Point Determination, https://www.youtube.com/watch?v=9RNRYLvlbXM.

C. E. Bell, D. Taber, and K. Clark, *Organic Chemistry Laboratory: Standard and Microscale Experiments*, 2nd Edition, Saunders College Publishing, San Diego (1997).

J. W. Zubrick, *The Organic Chem Lab Survival Guide*, John Wiley & Sons, Inc., Chichester (1988).

Melting Point Analysis protocol. Chem 211, Clark College (2007).

D. A. Straus, Melting Point Analysis, http://www.chemistry.sjsu.edu/straus/MP%20htms/MP1measure.htm.

Chapter 2: Molecular Weight Determination

Tawana Robinson, Sehmus Ozden, Yen-Hao Lin and
Andrew R. Barron

Background

The molecular mass (m) is the mass of a given molecule: it is measured in Daltons (Da or u). The definition of molecular weight (M_w) is synonymous with molecular mass; however, the units used are commonly g/mol. The terms molecular mass and molecular weight are often used interchangeably, however, the molecular mass is more commonly used when referring to the mass of a single or specific well-defined molecule and less commonly than molecular weight when referring to a weighted average of a sample. Thus, simple pure compounds contain the same molecular composition for the same species. For example, the molecular weight of any sample of styrene (Figure 2.1a) will be the same (104.16 g/mol). In contrast, most polymers, such as polystyrene (Figure 2.1b), are not composed of identical molecules. The molecular weight of a polymer is determined by the chemical structure of the monomer units, the lengths of the chains and the extent to which the chains are interconnected to form branched molecules. Because virtually all polymers are mixtures of many large molecules, we have to resort to averages to describe polymer molecular weight.

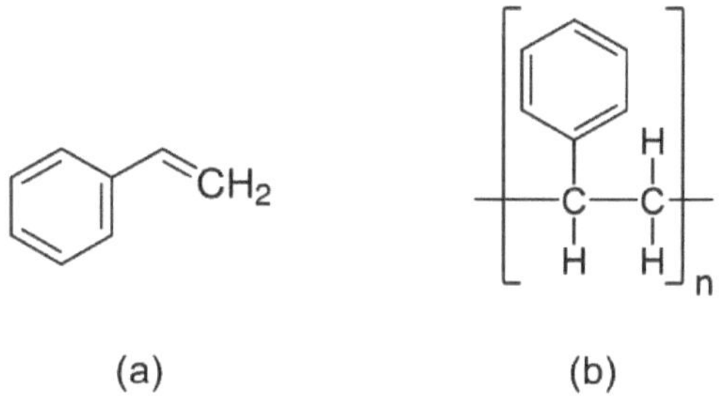

Figure 2.1: Molecular structure of (a) styrene and (b) polystyrene.

The polymers produced in polymerization reactions have lengths which are distributed according to a probability function which is governed by the polymerization reaction. To define a particular polymer weight average, the average molecular weight M_{avg} is defined by,

$$M_{avg} = \frac{\Sigma N_i M_i^a}{\Sigma N_i M_i^{a-1}}$$

where N_i is the number of molecules with molecular weight M_i.

There are several possible ways of reporting polymer molecular weight. Three commonly used molecular weight descriptions are: the number average (M_n), weight average (M_w), and z-average molecular weight (M_z). All of three are applicable to different constant *a* and are shown in Figure 2.2.

When a = 1, the number average molecular weight,

$$M_{n,avg} = \frac{\Sigma N_i M_i}{\Sigma N_i} = \frac{w}{N}$$

When a = 2, the weight average molecular weight,

$$M_{w,avg} = \frac{\Sigma N_i M_i^2}{\Sigma N_i M_i} = \frac{\Sigma N_i M_i}{w}$$

When a = 3, the z-average average molecular weight,

$$M_{z,avg} = \frac{\Sigma N_i M_i^3}{\Sigma N_i M_i^2} = \frac{\Sigma N_i M_i^2}{\Sigma N_i M_i}$$

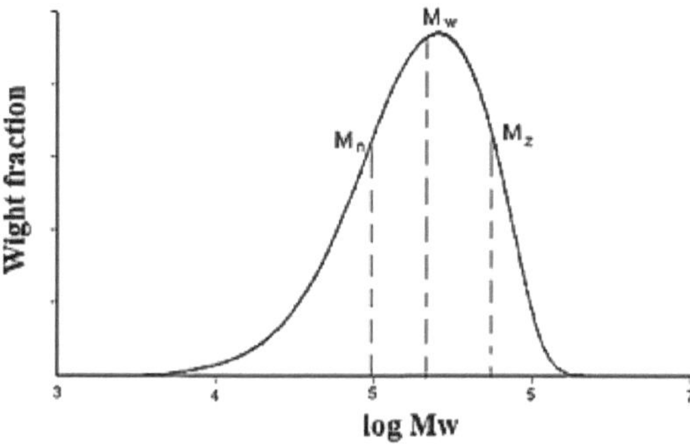

Figure 2.2: Distribution of molar masses for a polymer sample.

Bulk properties weight average molecular weight, M_w is the most useful one, because it fairly accounts for the contributions of different sized chains to the overall behavior of the polymer, and correlates best with most of the physical properties of interest.

There are various methods published to detect these three different primary average molecular weights respectively. For instance, a colligative method, such as osmotic pressure, effectively calculates the number of molecules present and provides a number average molecular weight regardless of their shape or size of polymers. The classical van't Hoff equation for the osmotic pressure of an ideal, dilute solution is,

$$\frac{\pi}{c} = \frac{RT}{M_n}$$

The weight average molecular weight of a polymer in solution can be determined by either measuring the intensity of light scattered by the solution or studying the sedimentation of the solute in an ultracentrifuge. From light scattering method which is depending on the size rather than the number of molecules, weight average molecular weight is obtained. This work requires concentration fluctuations which are the main source of the light scattered by a polymer solution. The intensity of the light scattering of polymer solution is often expressed by its turbidity τ which is given in Rayleigh's law,

$$\tau = \frac{16\pi i_\theta r^2}{3I_0(1+\cos2\theta)}$$

where i_θ is scattered intensity at only one angle θ, r is the distance from the scattering particle to the detection point, and I_0 is the incident intensity.

The intensity scattered by molecules (N_i) of molecular weight (M_i) is proportional to $N_iM_i^2$. Thus, the total light scattered by all molecules is described in,

$$\frac{\pi}{c} \sim \frac{\Sigma N_iM_i^2}{\Sigma N_iM_i}$$

where c is the total weight of the sample N_iM_i.

The viscosity of a polymer depends on concentration and molecular weight of polymers. Viscosity techniques is common since it is experimentally simple. Viscosity average molecular weight (M_v) defines as,

$$M_V = \left(\frac{\Sigma N_i M_i^{1+a}}{\Sigma N_i M_i} \right)^{1/a}$$

where M_i is molecular weight and N_i is number of molecules, a is a constant which depend on the polymer-solvent in the viscosity experiments. When a is equal to 1, M_v is equal to the weight average molecular weight, if it isn't equal 1 it is between weight average molecular weight and the number average molecular weight.

Distribution of molecular weight

Molecular weight distribution is one of the important characteristics of polymer because it affects polymer properties. A typical molecular distribution of polymers is shown in Figure 2.3. There is various molecular weight in the range of curve. The distribution of sizes in a polymer sample isn't totally defined by its central tendency. The width and shape of distribution must be known. It is always true that the various range molecular weight is,

$$M_N \leq M_V \leq M_W \leq M_Z \leq M_{Z+1}$$

The equality is occurring when all polymer in the sample have the same molecular weight.

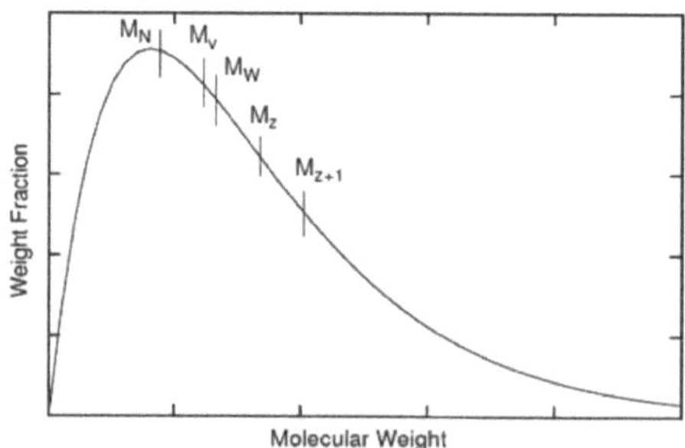

Figure 2.3: A schematic plot of a distribution of molecular weights along with the rankings of the various average molecular weights. Adapted from J. A. Nairn, Oregon State University (2003).

The polydispersity index (PDI), is a measure of the distribution of molecular mass in a given polymer sample. As shown in Figure 2.3, it is the result of the definitions that $M_w \geq M_n$. The equality of M_w and M_n would correspond with a perfectly uniform (monodisperse) sample. The ratio of these average molecular weights is often used as a guide to the dispersity of the chain lengths in a polymer sample. The greater M_w/M_n is, the greater the dispersity is.

The properties of a polymer sample are strongly dependent on the way in which the weights of the individual molecules are distributed about the average. The ratio M_w/M_n gives sufficient information to characterize the distribution when the mathematical form of the distribution curve is known.

Generally, the narrow molecular weight distribution materials are the models for much of work aimed at understanding the materials' behaviors. For example, polystyrene (Figure 2.1b) and its block copolymer polystyrene-b-polyisoprene (Figure 2.4) have quite narrow distribution. As a result, narrow molecular weight distribution materials are a necessary requirement when studying their behavior, such as self-assembly behavior for block copolymer. Nonetheless, there are still lots of questions for scientists to explore the influence of polydispersity. For example, research on self-assembly which is one of the interesting fields in polymer science shows that we cannot throw polydispersity away.

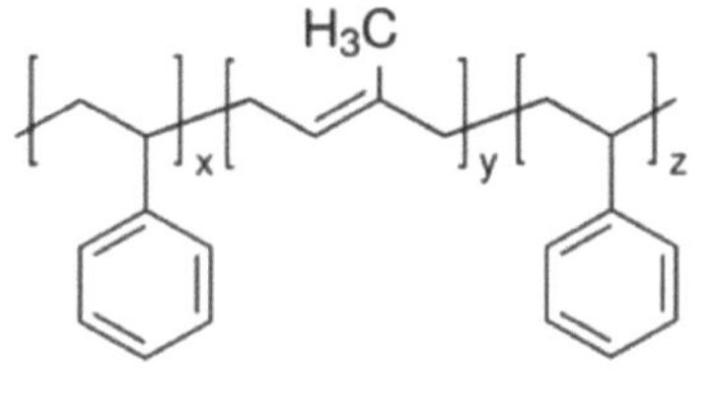

Figure 2.4: Structure of polystyrene-b-polyisoprene block copolymer.

Solution molecular weight of small molecules

The cryoscopic method was formally introduced in the 1880's when François-Marie Raoult (Figure 2.5) published how solutes depressed the freezing points of various solvents such as benzene, water, and formic acid. He concluded from his experimentation if one molecule of a substance can be dissolved in one-hundred molecules of any given solvent then the solvent temperature is lowered by a specific temperature increment. Based on Raoult's research, Ernst Otto Beckmann (Figure 2.6) invented the Beckmann thermometer and the associated freezing-point apparatus (Figure 2.7), which was a significant improvement in measuring freezing point depression values for a pure solvent. The simplicity, ease, and accuracy of this apparatus has allowed it to remain as a current standard with few modifications for molecular weight determination of unknown compounds.

Figure 2.5: French chemist François-Marie Raoult (1830 - 1901).

Figure 2.6: German chemist Ernst Otto Beckmann (1853 - 1923).

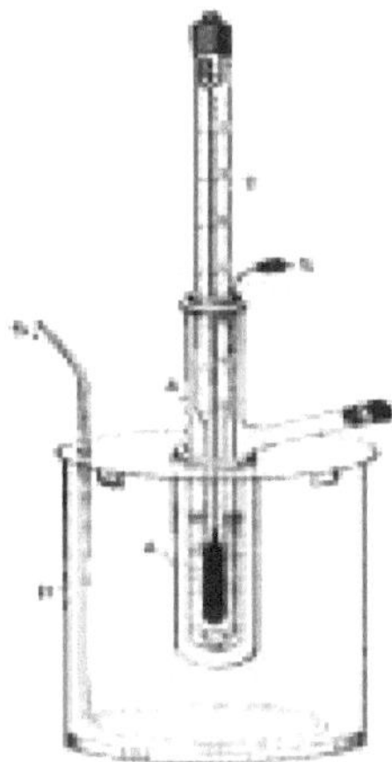

Figure 2.7: Beckmann differential thermometer and freezing point depression apparatus.

The historical significance of Raoult and Beckmann's research, among many other investigators, has revolutionized a physical chemistry technique that is currently applied to a vast range of disciplines from food science to petroleum fluids. For example, measured cryoscopic molecular weights of crude oil are used to predict the viscosity and surface tension for necessary fluid flow calculations in pipeline.

Freezing point depression

Freezing point depression is a colligative property in which the freezing temperature of a pure solvent decreases in proportion to the number of solute molecules dissolved in the solvent. The known mass of the added solute and the freezing point of the pure solvent information permit an accurate calculation of the molecular weight of the solute.

The freezing point depression of a non-ionic solution is described in,

$$\Delta T_f = K_f m$$

where ΔT_f is the change in the initial and final temperature of the pure solvent, K_f is the freezing point depression constant for the pure solvent, and m (moles solute/kg solvent) is the molality of the solution.

For an ionic solution the dissociation particles must be accounted for with the number of solute particles per formula unit, i (the van't Hoff factor),

$$\Delta T_f = K_f mi$$

Cryoscopic procedure

Cryoscopic apparatus

For cryoscopy, the apparatus to measure freezing point depression of a pure solvent may be representative of the Beckmann apparatus previously shown in Figure 2.7. The apparatus consists of a test tube containing the solute dissolved in a pure solvent, stir bar or magnetic wire and closed with a rubber stopper encasing a mercury thermometer. The test tube component is immersed in an ice-water bath in a beaker. An example of the apparatus is shown in Figure 2.8. The rubber stopper and stir bar/wire stirrer are not shown in Figure 2.8.

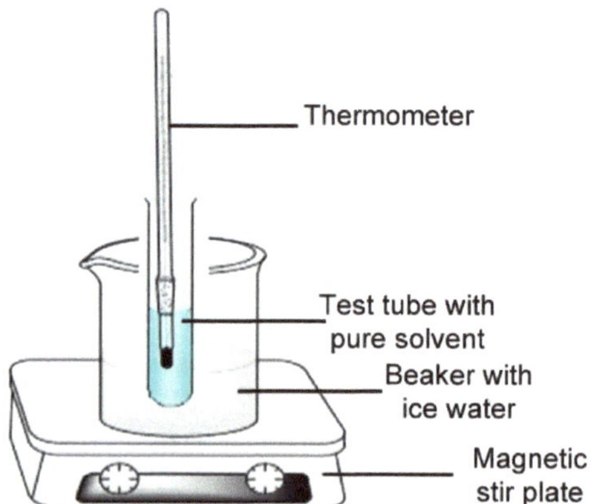

Figure 2.8: An example of a cryoscopic apparatus. Adapted from http://www.lahc.cc.ca.us/classes/chemistry/arias/Exp%2012%20-%20Freezing%20Point.pdf.

Sample and solvent selection

The cryoscopic method may be used for a wide range of samples with various degrees of polarity. The solute and solvent selection should follow the premise of like dissolved like or in terms of Raoult's principle of the dissolution of one molecule of solute in one-hundred molecules of a solvent. The most common solvents such as benzene are generally selected because it is unreactive, volatile, and miscible with many compounds. Table 2.1 shows the cryoscopic constants (K_f) for the common solvents used for cryoscopy. A complete list of K_f values are available in Knovel Critical Tables.

Compound	K_f
Acetic acid (CH_3CO_2H)	3.90
Benzene (C_6H_6)	5.12
Camphor ($C_{10}H_{16}O$)	39.7
Carbon disulfide (CS_2)	3.8
Carbon tetrachloride (CCl_4)	30
Chloroform ($CHCl_3$)	4.68
Cyclohexane (C_6H_{12})	20.2
Ethanol (C_2H_5OH)	1.99
Naphthalene ($C_{10}H_8$)	6.80
Phenol (C_6H_5OH)	7.27
Water	1.86

Table 2.1: Cryoscopic constants (K_f) for common solvents used for cryoscopy.

Cryoscopic Method

The detailed information about the procedure used for cryoscopy is shown below:

Step 1. Weigh (15 to 20 grams) of the pure solvent in a test tube and record the measured weight value of the pure solvent.

Step 2. Place a stir bar or wire stirrer in the test tube and close with a rubber stopper that has a hole to encase a mercury thermometer.

Step 3. Place a mercury thermometer in the rubber stopper hole.

Step 4. Immerse the test tube apparatus in an ice-water bath.

Step 5. Allow the solvent to stir continuously and equilibrate to a few degrees below the freezing point of the solvent.

Step 6. Record the temperature at which the solvent reaches the freezing point, which remains at a constant temperature reading.

Step 7. Repeat the freezing point data collection for at least two more measurements without a difference less than 0.5 °C between the measurements.

Step 8. Weigh a quantity of the solute for investigation and record the measured value.

Step 9. Add the weighed solute to the test tube containing the pure solvent.

Step 10. Re-close the test tube with a rubber stopper encasing a mercury thermometer.

Step 11. Re-immerse the test tube in an ice water bath and allow the mixture to stir to fully dissolve the solute in the pure solvent.

Step 12. Measure the freezing point and record the temperature value.

Allow the solution to stir continuously to avoid supercooling. The observed freezing point of the solution is when the temperature reading remains constant.

Sample calculation to determine molecular weight

Sample data set

Table 2.2 represents an example of a data set collection for cryoscopy.

Parameter	Trial 1	Trial 2	Trial 3	Average
Mass of cyclohexane (g)	9.05	9.00	9.04	9.03
Mass of unknown solute (g)	0.4000	0.4101	0.4050	0.4050
Freezing point of cyclohexane (°C)	6.5	6.5	6.5	6.5
Freezing point of cyclohexane mixed with unknown solute (°C)	4.2	4.3	4.2	4.2

Table 2.2: Example of a data set collection for cryoscopy.

Calculation of molecular weight using the freezing point depression equation

Calculate the freezing point (F_{pt}) depression of the solution, $T\Delta_f$ (from Table 2.2).

$$T\Delta_f = (Fpt \text{ of pure solvent}) - (Fpt \text{ of solution})$$

$T\Delta_f = 6.5\,°C - 4.2\,°C$

$T\Delta_f = 2.3\,°C$

Calculate the molal concentration, m, of the solution using the freezing point depression and K_f (see Table 2.1 and Table 2.2).

$T\Delta_f = K_f m$

$m = (2.3\,°C)/(20.2\,°C/molal)$

$m = 0.113\ molal$

Note that,

$m = g\ (solute)/kg\ (solvent)$

Calculate the M_w of the unknown sample. Note that for i = 1 for covalent compounds in,

$$M_W = \frac{K_f\ (g\ solute)}{\Delta T_f\ (kg\ solvent)}$$

$$M_W = \frac{20.2\,°C*kg/moles \times 0.405\ g}{2.3\,°C \times 0.00903\ kg}$$

$M_W = 393\ g/mol$

Molecular weight of polymers

Knowledge of the molecular weight of polymers is very important because the physical properties of macro-molecules are affected by their molecular weight. For example, shown in Figure 2.9 the interrelation between molecular weight and strength for a typical polymer.

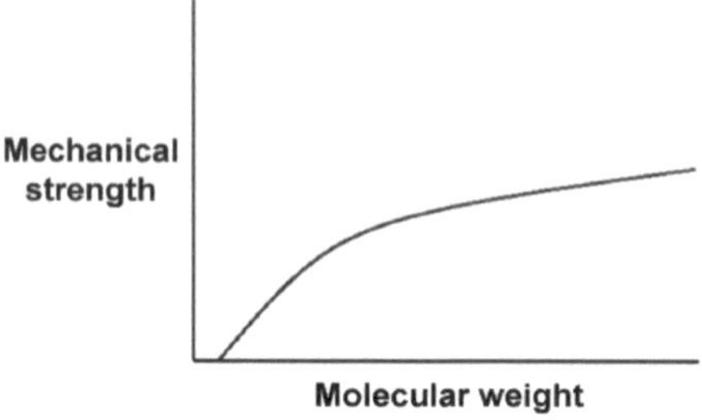

Figure 2.9: Dependence of mechanical strength on polymer molecular weight. Adapted from G. Odian, *Principles of Polymerization*, 4th edn., Willey-Interscience, New York (2004).

The melting point of polymers are also slightly dependent on their molecular weight. (Figure 2.10) shows relationship between molecular weight and melting temperatures of polyethylene (Figure 2.11). Most linear low-density polyethylenes (Figure 2.12) have melting temperatures near 140 °C. The approach to the theoretical asymptote, that is a line whose distance to a given curve tends to zero, indicative that a theoretical polyethylene of infinite molecular weight (i.e., M = ∞) would have a melting point of 145 °C.

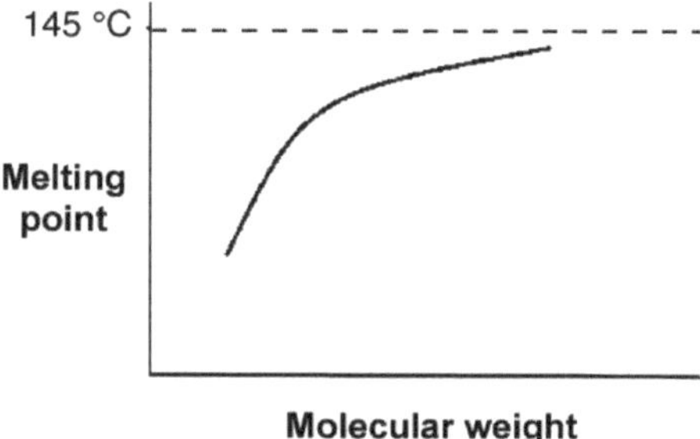

Figure 2.10: The molecular weight-melting temperature relationship for the alkane series. Adapted from L. H. Sperling, *Introduction to Physical Polymer Science*, 4th edn., Wiley-Interscience, New York (2005).

$$\left(\begin{array}{ccc} & H & H \\ & | & | \\ - & C - C & - \\ & | & | \\ & H & H \end{array}\right)_n$$

Figure 2.11: Structure of polyethylene.

Figure 2.12: Linear low-density polyethylene beads.

Other properties that the molecular weight (M_w) and molecular weight distribution influences include:

- **Processability:** the suitability of the polymer to physical processing.
- **Glass-transition temperature:** refers to the transformation of a glass-forming liquid into a glass.
- **Solution viscosity:** measure of the resistance of a fluid which is being deformed by either shear stress or tensile stress.
- **Hardness:** a measure of how resistant a polymer is to various kinds of permanent shape change when a force is applied.
- **Melt viscosity:** the rate of extrusion of thermoplastics through an orifice at a prescribed temperature and load.
- **Tear strength:** a measure of the polymers resistance to tearing.
- **Tensile strength:** as indicated by the maxima of a stress-strain curve and, in general, is the point when necking occurs upon stretching a sample.
- **Stress-crack resistance:** the formation of cracks in a polymer caused by relatively low tensile stress and environmental conditions.
- **Brittleness:** the liability of a polymer to fracture when subjected to stress.
- **Impact resistance:** the relative susceptibility of polymers to fracture under stresses applied at high speeds.
- **Flex life:** the number of cycles required to produce a specified failure in a specimen flexed in a prescribed manner.
- **Stress relaxation:** describes how polymers relieve stress under constant strain.
- **Toughness:** the resistance to fracture of a polymer when stressed.

- **Creep strain:** the tendency of a polymer to slowly move or deform permanently under the in fluence of stresses.
- **Drawability:** The ability of fiber-forming polymers to undergo several hundred percent permanent deformations under load, at ambient or elevated temperatures.
- **Compression:** the result of the subjection of a polymer to compressive stress.
- **Fatigue:** the failure by repeated stress.
- **Tackiness:** the property of a polymer being adhesive or gummy to the touch.
- **Wear:** the erosion of material from the polymer by the action of another surface.
- **Gas permeability:** the permeability of gas through the polymer.

Consequently, it is important to understand how to determine the molecular weight and molecular weight distribution.

Gel permeation chromatography (GPC)/size exclusion chromatography (SEC)

Gel permeation chromatography is also called size exclusion chromatography. It is widely used method to determine high molecular weight distribution. In this technique, substances separate according to their molecule size. Firstly, large molecules begin to elute then smaller molecules are eluted Figure 2.13. The sample is injected into the mobile phase then the mobile phase enters into the columns. Retention time is the length of time that a particular fraction remains in the column. As shown in Figure 2.13, while the mobile phase passes through the porous particles, the area between large molecules and small molecules is getting increase. GPC gives a full molecular distribution, but its cost is high.

According to basic theory of GPC, the basic quantity measured in chromatography is the retention volume,

$$V_e = V_0 + V_p K$$

where V_0 is mobile phase volume and V_p is the volume of a stationary phase. K is a distribution coefficient related to the size and types of the molecules.

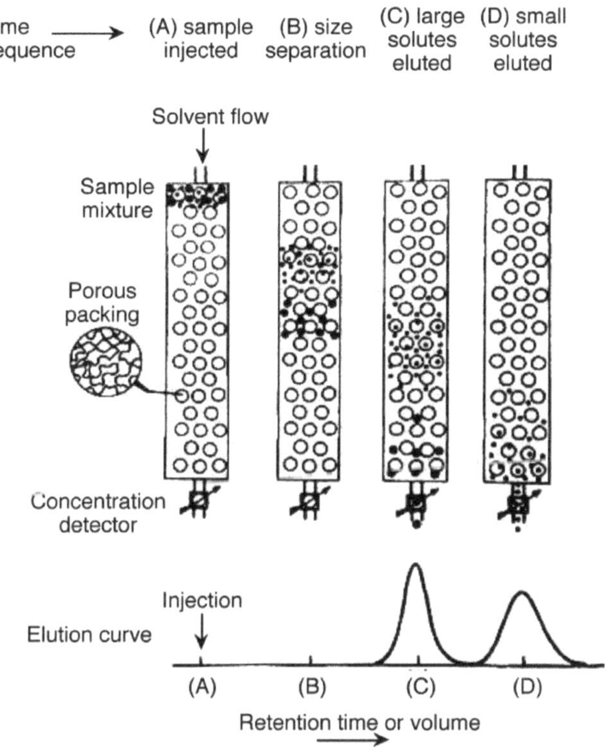

Figure 2.13: Solvent flow through column. Adapted from A. M. Striegel, W. W. Yau, J. J. Kirkland, and D. D. Bly. *Modern Size-Exclusion Liquid Chromatography - Practice of Gel Permeation and Gel Filtration Chromatography*, 2nd edn., Hoboken. NJ (2009).

The essential features of gel permeation chromatography are shown in Figure 2.14. Solvent leaves the solvent supply, then solvent is pumped through a filter. The desired amount of flow through the sample column is adjusted by sample control valves and the reference flow is adjusted that the flow through the reference and flow through the sample column reach the detector in common front. The reference column is used to remove any slight impurities in the solvent. In order to determine the amount of sample, a detector is located at the end of the column. Also, detectors may be used to continuously verify the molecular weight of species eluting from the column. The flow of solvent volume is as well monitored to provide a means of characterizing the molecular size of the eluting species. Figure 2.15 shows the regular instrumental setup of SEC.

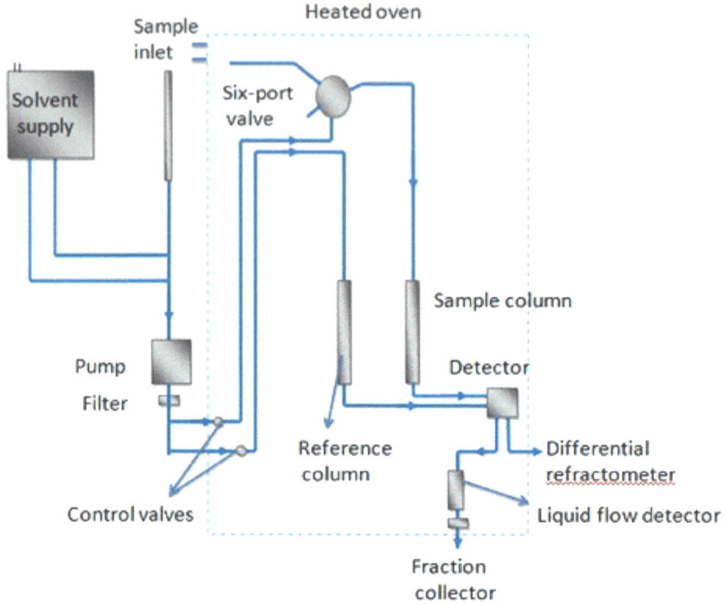

Figure 2.14: Schematic of gel permeation chromatography system.

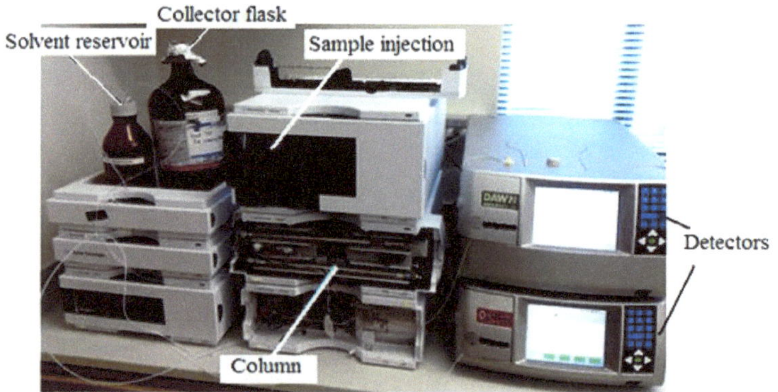

Figure 2.15: Regular instrumentation for GPC.

Solvent selection

Solvent selection for GPC involves a number if considerations, such as convenience, sample type, column packing, operating variables, safety, and purity.

For samples concern, the solvents used for mobile phase of GPC are limited to those follows following criteria:

- The solvent must dissolve the sample completely.
- The solvent has different properties with solute in the eluent: typically, with solvent refractive index (RI) different from the sample RI by ±0.05 unit of more, or more than 10% of incident energy for UV detector.
- The solvent must not degrade the sample during use. Otherwise, the viscosity of eluent will gradually increase over times.
- The solvent is not corrosive to any components of the equipment.

Therefore, several solvents are qualified to be solvents such as THF, chlorinated hydrocarbons (chloroform, methylene chloride, dichloroethane, etc.), aromatic hydrocarbons (benzene, toluene, trichlorobenzene, etc.). Normally, high purity of solvent (HPLC-grade) is recommended. The reasons are to avoid suspended particulates that may abrade the solvent pumping system or cause plugging of small-particle columns, to avoid impurities that may generate baseline noise, and to avoid impurities that are concentrated due to evaporation of solvent.

Column selection

Column selection of SEC depends mainly on the desired molecular weight range of separation and the nature of the solvents and samples. Solute molecules should be separated solely by the size of gels without interaction of packing materials. Better column efficiencies and separations can be obtained with small particle packing in columns and high diffusion rates for sample solutes. Furthermore, optimal performance of an SEC packing materials involves high resolution and low column backpressure. Compatible solvent and column must be chosen because, for example, organic solvent is used to swell the organic column packing and used to dissolve and separate the samples.

It is convenient that columns are now usually available from manufacturers regarding the various types of samples. They provide the information such as maximum tolerant flow rates, backpressure tolerances, recommended sample concentration, and injection volumes, etc. Nonetheless, users have to notice a few things upon using columns:

Vibration and extreme temperatures should be avoided because these will post irreversible damage on columns.

For aqueous mobile phase, it is unwise to allow the extreme pH solutions staying in the columns for a long period of time.

The columns should be stored with some neat organic mobile phase, or aqueous mobile phase with pH range 2 - 8 to prevent degradation of packing when not in use.

Sample preparation

The sample solutions are supposed to be prepared in dilute concentration (less than 2 mg/mL) for several concerns. For polymer samples, samples must be dissolved in the solvent same as used for mobile phase except some special cases. A good solvent can dissolve a sample in any proportion in a range of temperatures. It is a slow process for dissolution because the rate determining step is solvent diffusion into polymers to produce swollen gels. Then, gradual disintegration of gels makes sample-solvent mixture truly become solution. Agitation and warming the mixtures are useful methods to speed up sample preparation.

It is recommended to filter the sample solutions before injecting into columns or storing in sample vials in order to get rid of clogging and excessively high-pressure problems. If unfortunately, the excessively high pressure or clogging occur due to higher concentration of sample solution, raising the column temperature will reduce the viscosity of the mobile phase, and may be helpful to re-dissolve the precipitated or adsorbed solutes in the column. Back flushing of the columns should only be used as the last resort.

Analysis of GPC data

The size exclusion separation mechanism is based on the effective hydrodynamic volume of the molecule, not the molecular weight, and therefore the system must be calibrated using standards of known molecular weight and homogeneous chemical composition. Then, the curve of sample is used to compare with calibration curve and obtain information relative to standards. The further step is required to covert relative molecular weight into absolute molecular weight of a polymer.

Calibration

The purpose of calibration in SEC is to de ne the relationship between molecular weight and retention volume/time in the chosen permeation range of column set and to calculate the relative molecular weight to standard molecules. There are several calibration methods are commonly employed in modern SEC: direct standard calibration, poly-disperse standard calibration, universal calibration.

The most commonly used calibration method is direct standard calibration. In the direct standard calibration method, narrowly distributed standards of the same polymer being analyzed are used. Normally, narrow-molecular weight standards commercially available are polystyrene (PS). The molecular weight of standards is measured originally by membrane osmometry for number-average molecular weight, and by light scattering for weight-average molecular weight as described above. The retention volume at the peak maximum of each standard is equated with its stated molecular weight (Figure 2.16).

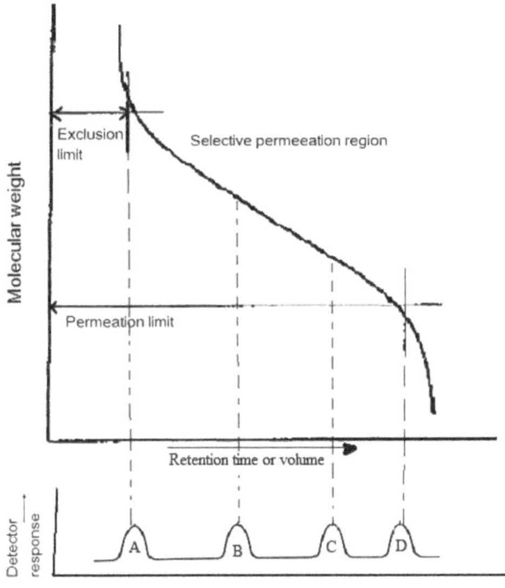

Figure 2.16: Calibration curve for a size-exclusion.

Relative M_w versus absolute M_w

The molecular weight and molecular weight distribution can be determined from the calibration curves as described above. But as the relationship between molecular weight and size depends on the type of polymer, the calibration curve depends on the polymer used, with the result that true molecular weight can only be obtained when the sample is the same type as calibration standards. As Figure 2.17 depicted, large deviations from the true molecular weight occur in the instance of branched samples because the molecular density of these is higher than in the linear chains.

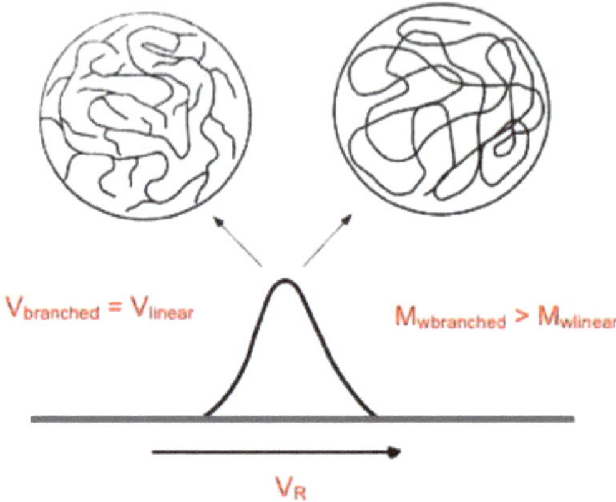

Figure 2.17: GPC elution of linear and branched samples of similar hydrodynamic volumes, but different molecular weights. Adapted from S. Mori, and H. G. Barth. *Size Exclusion Chromatography*, **Springer, New York. (1999). Copyright: Springer (1999).**

Light-scattering detectors are now often used to overcome the limitations of conventional GPC. These signals depend only on concentration, not on molecular weight or polymer size. For instance, for LS detector applies,

$$\text{LS signal} = K_{LS}(dn/dc)^2 M_w c$$

where K_{LS} is an apparatus-specific sensitivity constant, dn/dc is the refractive index increment and c is concentration. Therefore, accurate molecular weight can be determined while the concentration of the sample is known without calibration curve.

Examples

As an example, consider the block copolymer of ethylene glycol (PEG, Figure 2.18) and poly(lactide) (PLA, Figure 2.19), i.e., Figure 2.20. In the first step starting with a sample of PEG with a M_n of 5,700 g/mol. After polymerization, the molecular weight increased because of the progress of lactide polymerization initiated from end of PEG chain. Varying composition of PEG-PLA shown in Table 2.3 can be detected by GPC (Figure 2.21).

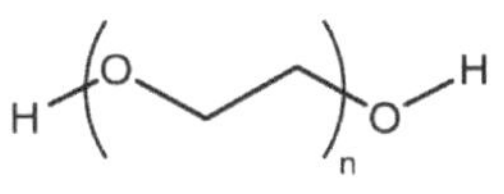

Figure 2.18: The structure of polyethyleneglycol (PEG).

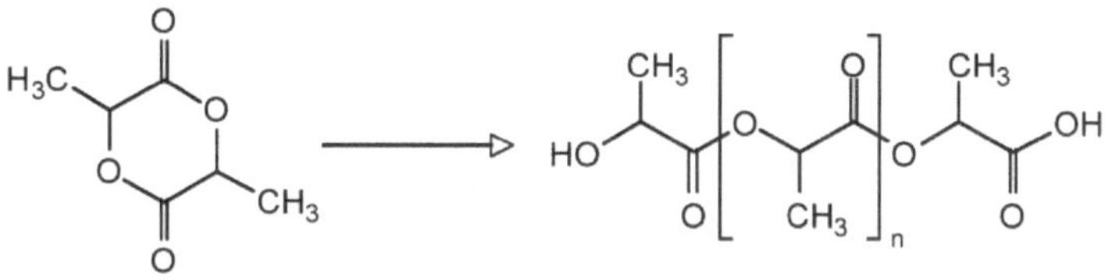

Figure 2.19: The ring-opening polymerization of lactide to polylactide.

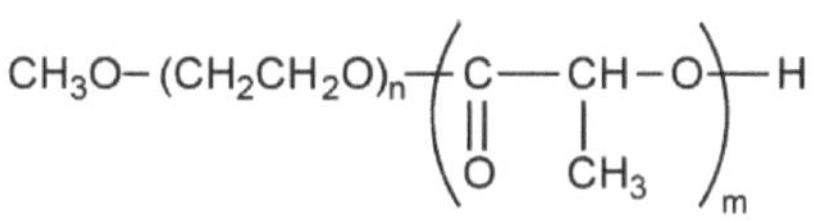

Figure 2.20: The structure of PEG-PLA block copolymer.

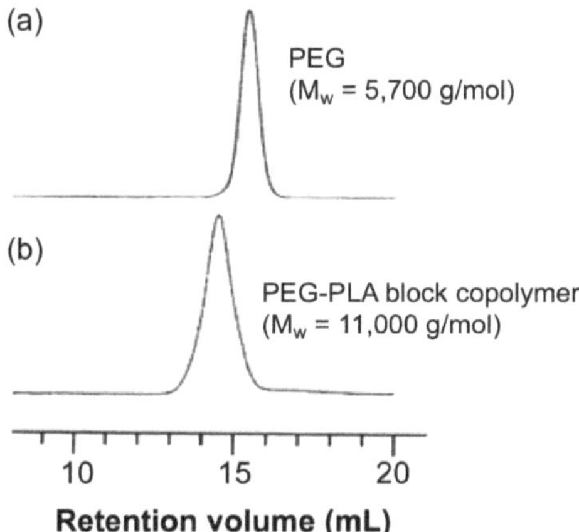

Figure 2.21: Gel permeation chromatogram of (a) PEG (M_w = 5,700 g/mol) and (b) PEG-PLA block copolymer (M_w = 11,000 g/mol). Adapted from K. Yasugi, Y. Nagasaki, M. Kato, K. Kataoka, Preparation and characterization of polymer micelles from poly(ethylene glycol)-poly(d,l-lactide) block copolymers as potential drug carrier. *J. Control. Release*, 1999, 62, 89. Copyright: Elsevier (1999).

Polymer	M_n of PEG	M_w/M_n of PEG	M_n of PLA	M_w/M_n of block copolymer	Weight ratio of PLA to PEG
PEG-PLA(41-12)	4100	1.05	1200	1.05	0.29
PEG-PLA(60-30)	6000	1.03	3000	1.08	0.50
PEG-PLA(57-54)	5700	1.03	5400	1.08	0.95
PEG-PLA(61-78)	6100	1.03	7800	1.11	1.28

Table 2.3: Characteristics of PEG-PLA block copolymer with varying composition. Data from K. Yasugi, Y. Nagasaki, M. Kato, K. Kataoka, Preparation and characterization of polymer micelles from poly(ethylene glycol)-poly(d,l-lactide) block copolymers as potential drug carrier. *J. Control. Release*, **1999, 62, 89.**

The syntheses of poly(3-hexylthiophene) are well developed during last decade. It is an attractive polymer due to its potential as electronic materials. Due to its excellent charge transport performances and high solubility, several studies discuss its further improvement such as making block copolymer even triblock copolymer. The details are not discussed here. However, the importance of molecular weight and molecular weight distribution is still critical.

As shown in Figure 2.23, the mechanism of chain-growth polymerization was studied and successfully produced low polydispersity P3HT. The figure also demonstrates that the molecule with larger molecular size/ or weight elutes out of the column earlier than those which has smaller molecular weight (Figure 2.24).

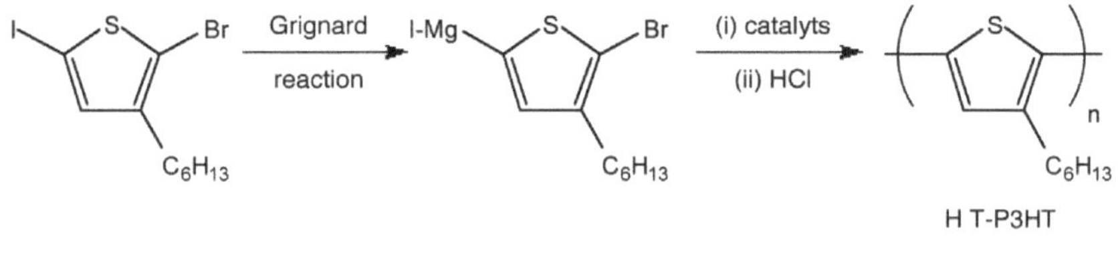

Figure 2.23: Synthesis of a well-defined poly(3-hexylthiphene) (HT-P3HT).

The real molecular weight of P3HT is smaller than the molecular weight relative to polystyrene. In this case, the backbone of P3HT is harder compared with polystyrenes' backbone because of the position of aromatic groups. It results in less flexibility. We can briefly judge the authentic molecular weight of the synthetic polymer according to its molecular structure.

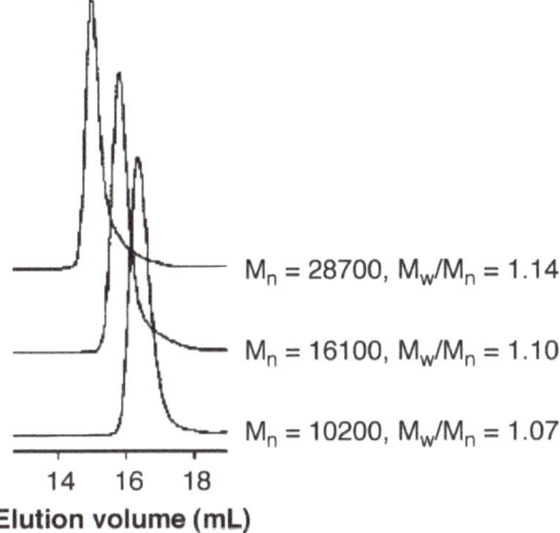

Figure 2.24: GPC profiles of HT-P3HT obtained by the polymerization. Adapted from R. Miyakoshi, A. Yokoyama, and T. Yokozawa, Synthesis of poly(3-hexylthiophene) with a narrower polydispersity. *Macromol. Rapid Commun.*, 2004, 25, 1663. Copyright: Wiley-VCH (2004).

Light-scattering

One of the most used methods to characterize the molecular weight is light scattering method. When polarizable particles are placed in the oscillating electric field of a beam of light, the light scattering occurs. Light scattering method depends on the light, when the light is passing through polymer solution, it is measure by loses energy because of absorption, conversion to heat and scattering. The intensity of scattered light relies on the concentration, size and polarizability that is proportionality constant which depends on the molecular weight. Figure 2.25 shows light scattering of a particle in solution.

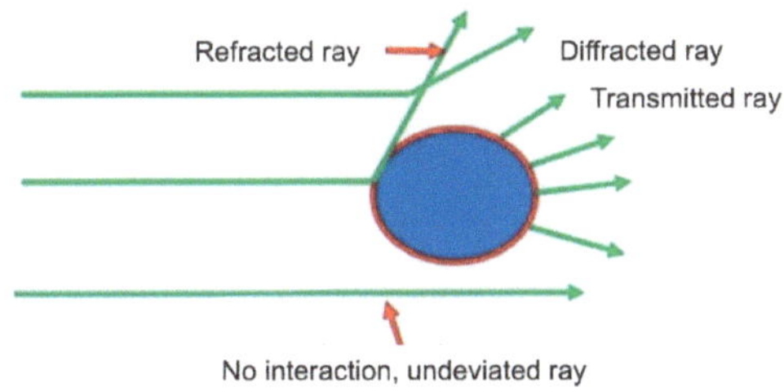

Figure 2.25: Modes of scattering of light in solution.

A schematic laser light-scattering is shown in Figure 2.26. A major problem of light scattering is to prepare perfectly clear solutions. This problem is usually accomplished by ultra-centrifugation. A solution should be as possible as clear and dust free to determine absolute molecular weight of polymer. The advantages of this method, it doesn't need calibration to obtain absolute molecular weight and it can give information about shape and M_w information. Also, it can be performed rapidly with less amount of sample and absolute determinations of the molecular weight can be measured. The weaknesses of the method are high price and most times it requires difficult clarification of the solutions.

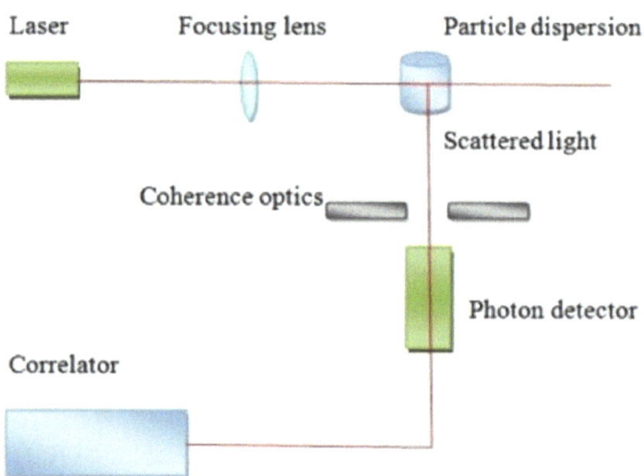

Figure 2.26: Schematic representation of light scattering.

The weight average molecular weight value of scattering polymers in solution related to their light scattering properties are defined by,

$$KC/R(\theta) = 1/ M_w(P(\theta) + 2A_2C + 3A_3C_2 + \ldots)$$

where K is the wave vector, that is defined by,

$$K = 2\pi^2 n_0^2 (dn/dC)^2/N_a\lambda^4$$

where C is solution concentration, $R(\theta)$ is the reduced Rayleigh ratio, $P(\theta)$ the particle scattering function, θ is the scattering angle, A is the osmotic virial coefficients, where n_0 solvent refractive index, λ the light wavelength and N_a

is Avogadro's number (6.022×10^{23}). The particle scattering function is given by,

$$1/(P(\theta) = 1 + 16\pi^2 n_0^2 (R_z^2) \sin^2(\theta/2) 3\lambda^2$$

where R_z is the radius of gyration.

Weight average molecular weight of a polymer is found from extrapolation of data in the form of a Zimm plot (Figure 2.27). Experiments are performed at several angles and at least at 4 different concentrations. The straight-line extrapolations provide M_w.

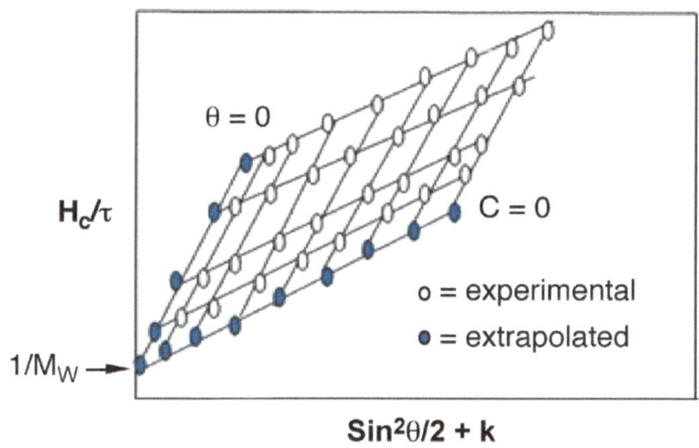

Figure 2.27: A typical Zimm plot of light scattering data. Adapted from M. P. Stevens, *Polymer Chemistry an Introduction*, 3rd edn., Oxford University Press, Oxford (1999).

X-ray scattering

X-rays are a form of electromagnetic wave with wavelengths between 0.001 nm and 0.2 nm. X-ray scattering is particularly used for semi-crystalline polymers which includes thermoplastics, thermoplastic elastomers, and liquid crystalline polymers. Two types of X-ray scattering are used for polymer studies:

- Wide-angle X-ray scattering (WAXS) which is used to study orientation of the crystals and the packing of the chains.

- Small-angle X-ray scattering (SAXS) which is used to study the electron density fluctuations that occur over larger distances as a result of structural inhomogeneities.

Schematic representation of X-ray scattering shows in Figure 2.28.

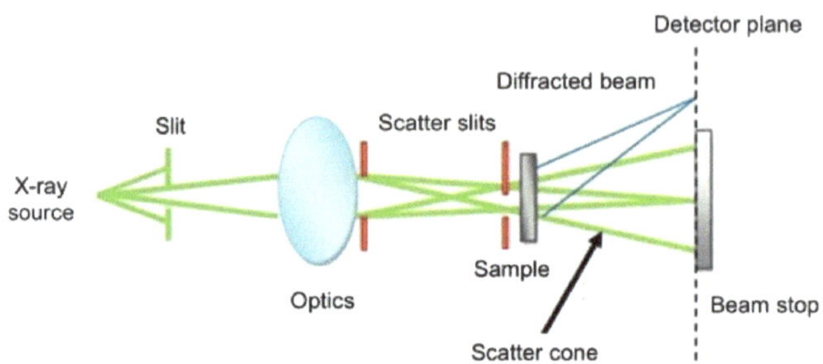

Figure 2.28: Schematic diagram of X-ray scattering. Adapted from B. Chu, and B. S. Hsiao, Small-angle X-ray scattering of polymers. *Chem. Rev.*, 2001, 101, 1727. Copyright: American Chemical Society (2001).

At least two SAXS curves are required to determine the molecular weight of a polymer. The SAXS procedure to determine the molecular weight of polymer sample in monomeric or multimeric state solution requires the following conditions.
- The system should be monodispersed.
- The solution should be dilute enough to avoid spatial correlation effects.
- The solution should be isotropic.
- The polymer should be homogeneous.

Osmometer

Osmometry is applied to determine number average of molecular weight (M_n). There are two types of osmometer:
- Vapor pressure osmometry (VPO).
- Membrane osmometry.

Vapor pressure osmometry measures vapor pressure indirectly by measuring the change in temperature of a polymer solution on dilution by solvent vapor and is generally useful for polymers with M_n below 10,000 - 40,000 g/mol. When molecular weight is more than that limit, the quantity being measured becomes very small to detect. A typical vapor osmometry shows in the Figure 2.29. Vapor pressure is very sensitive because of this reason it is measured indirectly by using thermistors to measure voltage changes caused by changes in temperature.

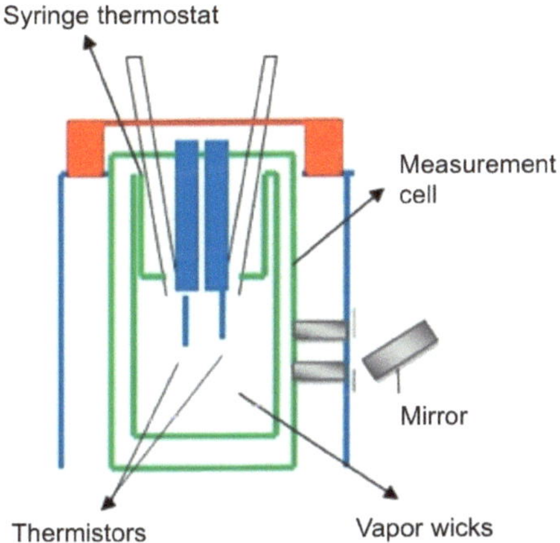

Figure 2.29: Schematic vapor pressure osmometry.

Membrane osmometry is absolute technique to determine M_n (Figure 2.30). The solvent is separated from the polymer solution with semipermeable membrane that is strongly held between the two chambers. One chamber is sealed by a valve with a transducer attached to a thin stainless-steel diaphragm which permits the measurement of pressure in the chamber continuously. Membrane osmometry is useful to determine M_n about 20,000 - 30,000 g/mol and less than 500,000 g/mol. When M_n of polymer sample more than 500,000 g/mol, the osmotic pressure of polymer solution becomes very small to measure absolute number average of molecular weight. In this technique, there are problems with membrane leakage and symmetry. The advantages of this technique are that it doesn't require calibration and it gives an absolute value of M_n for polymer samples.

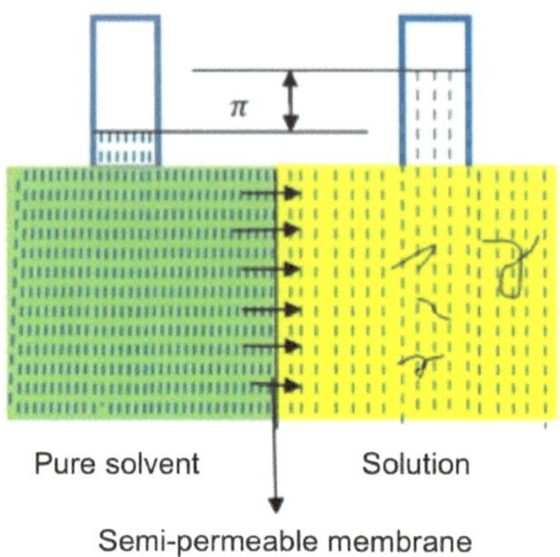

Figure 2.30: Schematic representative of membrane osmometry.

Summary

Properties of polymers depend on their molecular weight. There are different kind of molecular weight and each can be measured by different techniques. The summary of these techniques and molecular weight is shown in the Table 2.4.

Method	Type of molecular weight	Range of application
Light scattering	M_w	∞
Membrane osmometry	M_n	$10^4 - 10^6$
Vapor phase osmometry	M_n	40,000
X-ray scattering	M_w, M_n, M_z	$10^2 - 10^6$

Table 2.4: Summary of techniques of molecular weight of polymers.

Bibliography

H. Allcock, and F. W. Lampe. *Contemporary Polymer Chemistry*, 2nd edn., Englewood Cliffs, NJ (1990).

E. Beckmann and P. E. Tig, Kryoskopische bestimmungen bei tiefen temperaturen (−40 bis −117°). *Z. Anorg. Allg. Chem.*, 1910, **67**, 17.

B. H. Bersted, Molecular weight determination of high polymers by means of vapor pressure osmometry and the solute dependence of the constant of calibration. *J. Appl. Polym. Sci.*, 1973, **17**, 1415.

B. Chu and B. S. Hsiao, Small-angle X-ray scattering of polymers. *Chem. Rev.*, 2001, **101**, 1727.

A. H. Compton, On the mechanism of X-ray scattering. *Physics*, 1925, **11**, 303.

J. M. G. Cowie, *Polymers: Chemistry and Physics of Modern Materials*, 3rd edn., CRC Press (2007).

V. E. Eskin, Light scattering as a method of studying polymers. *Sov. Phys. Usp.*, 1964, **7**, 270.

P. J. Flory, *Principle of Polymer Chemistry*, Cornell University Press, Ithaca (1953).

M. A. Hillmyer, Polydisperse block copolymers: Don't throw them away. *J. Polymer Sci. B: Polymer Phys.*, 2007, **45**, 3249.

A. Horta and M. A. Pastoriza, The molecular weight distribution of polymer samples. *J. Chem. Educ.*, 2007, **84**, 1217.

C. H. Hsu, P. M. Peacock, R. B. Flippen, S. K. Manohar, and A. G. MacDiarmid, The molecular weight of polyaniline by light scattering and gel permeation chromatography. *Synth. Met.*, 1993, **60**, 233.

H. C. Jones, *The Modern Theory of Solution: Memoirs by Pfeffer, Van't Hoff, Arrhenius, and Raoult - Primary Source Edition*, Harper and Brothers, New York (1899).

Knovel Critical Tables, 2nd Ed., Knovel, New York (2008).

G. Odian, *Principle of Polymerization*, 4th edn., Willey Intersicence, New York (2004).

D. A McQuarrie, P. A Rock, and E. B. Gallogly, *General Chemistry*, University Science Books, Mill Valley (2011).

P. Meares, *Polymers: Structure and Bulk Properties*, Van Nostrand, New York (1971).

J. Mitchell. *Applied Polymer Analysis and Characterization- Recent Developments in Techniques, Instrumentation, Problem Solving*, Verlag, Munich (1987).

R. Miyakoshi, A. Yokoyama, and T. Yokozawa, Synthesis of poly(3-hexylthiophene) with a narrower polydispersity. *Macromol. Rapid Commun.*, 2004, **25**, 1663.

R. Miyakoshi, A. Yokoyama, and T. Yokozawa, Catalyst-transfer polycondensation. mechanism of Ni-catalyzed chain-growth polymerization leading to well-defined poly(3-hexylthiophene). *J. Am. Chem. Soc.*, 2005, **127**, 17542.

S. Mori, and H. G. Barth, *Size Exclusion Chromatography*, Springer, Berlin (1999).

R. W. Nunes, J. R. Martin, and J. F. Johnson, Influence of molecular weight and molecular weight distribution on mechanical properties of polymers. *Poly. Eng. Sci.*, 1982, **22**, 4.

H. P. Ronningsen, Prediction of viscosity and surface tension of North Sea petroleum fluids by using the average molecular weight. *Energy Fuels*, 1993, 7, 565.

L. H. Sperling, *Introduction to physical polymer science*, 4th edn., Wiley-Interscience, New York (2006).

M. P. Stevens, *Polymer Chemistry an Introduction*, 3rd edn., Oxford University Press, Oxford (1998).

A. M. Striegel, W. W. Yau, J. J. Kirkland, and D. D. Bly. *Modern Size-Exclusion Liquid Chromatography-Practice of Gel Permeation and Gel Filtration Chromatography*, 2nd edn., Hoboken, NJ (2009).

T. Tanaka, *Experimental Methods in Polymer Science*, Academic Press, New York (1999).

K. Yasugi, Y. Nagasaki, M. Kato, and K. Kataoka, Preparation and characterization of polymer micelles from poly(ethylene glycol)-poly(d,l-lactide) block copolymers as potential drug carrier. *J. Control. Release*, 1999, **62**, 89.

Chapter 3: BET Surface Area Analysis

Nina Hwang and Andrew R. Barron

Introduction

In the past few years, nanotechnology research has expanded out of the chemistry department and into the fields of medicine, energy, aerospace and even computing and information technology. With bulk materials, the surface area to volume is insignificant in relation to the number of atoms in the bulk, however when the particles are only 1 to 100 nm across, different properties begin to arise. For example, commercial grade zinc oxide has a surface area range of 2.5 - 12 m^2/g while nanoparticle zinc oxide can have surface areas as high as 54 m^2/g. The nanoparticles have superior UV blocking properties when compared to the bulk material, making them useful in applications such as sunscreen. Many useful properties of nanoparticles rise from their small size, making it very important to be able to determine their surface area.

Overview of BET theory

The BET theory was developed by Stephen Brunauer (Figure 3.1), Paul Emmett (Figure 3.2), and Edward Teller (Figure 3.3) in 1938. The first letter of each publisher's surname was taken to name this theory. The BET theory was an extension of the Langmuir theory, developed by Irving Langmuir (Figure 3.4) in 1916.

Figure 3.1: Hungarian chemist Stephen Brunauer (1903 - 1986).

Figure 2.2: American chemical engineer Paul H. Emmett (1900 - 1985).

Figure 2.3: Hungarian born theoretical physicist Edward Teller (1908 2003) shown in 1958 as the director of Lawrence Livermore National Laboratory was known as "the father of the hydrogen bomb".

Figure 3.4: American chemist and physicist Irving Langmuir (1881 - 1957).

The Langmuir theory relates the monolayer adsorption of gas molecules (Figure 3.5), also called adsorbates, onto a solid surface to the gas pressure of a medium above the solid surface at a fixed temperature to,

$$\theta = \frac{\alpha \cdot P}{1 + (\alpha \cdot P)}$$

where θ is the fractional cover of the surface, P is the gas pressure and α is a constant.

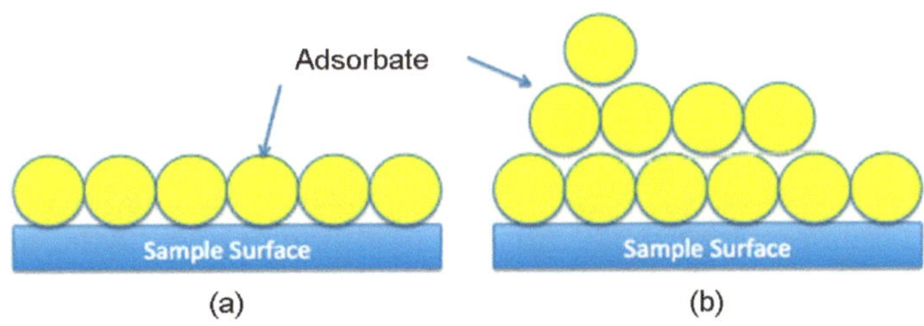

Figure 3.5: Schematic of the adsorption of gas molecules onto the surface of a sample showing (a) the monolayer adsorption model assumed by the Langmuir theory and (b) is the multilayer adsorption model assumed by the BET theory.

The Langmuir theory is based on the following assumptions:
- All surface sites have the same adsorption energy for the adsorbate, which is usually argon, krypton or nitrogen gas. The surface site is defined as the area on the sample where one molecule can adsorb onto.
- Adsorption of the solvent at one site occurs independently of adsorption at neighboring sites.
- Activity of adsorbate is directly proportional to its concentration.
- Adsorbates form a monolayer.
- Each active site can be occupied only by one particle.

The Langmuir theory has a few flaws that are addressed by the BET theory. The BET theory extends the Langmuir theory to multilayer adsorption (Figure 3.5) with three additional assumptions:
- Gas molecules will physically adsorb on a solid in layers in finitely.
- The different adsorption layers do not interact.
- The theory can be applied to each layer.

How does BET work?

Adsorption is defined as the adhesion of atoms or molecules of gas to a surface. It should be noted that adsorption is not confused with absorption, in which a fluid permeates a liquid or solid. The amount of gas adsorbed depends on the exposed surface are but also on the temperature, gas pressure and strength of interaction between the gas and solid. In BET surface area analysis, nitrogen is usually used because of its availability in high purity and its strong interaction with most solids. Because the interaction between gaseous and solid phases is usually weak, the surface is cooled using liquid N_2 to obtain detectable amounts of adsorption. Known amounts of nitrogen gas are then released stepwise into the sample cell. Relative pressures less than atmospheric pressure are achieved by creating conditions of partial vacuum. After the saturation pressure, no more adsorption occurs regardless of any further increase in pressure. Highly precise and accurate pressure transducers monitor the pressure changes due to the adsorption process. After the adsorption layers are formed, the sample is removed from the nitrogen atmosphere and heated to cause the adsorbed nitrogen to be released from the material and quantified. The data collected is displayed in the form of a BET isotherm, which plots the amount of gas adsorbed as a function of the relative pressure. There are five types of adsorption isotherms possible.

Type I isotherm

Type I is a pseudo-Langmuir isotherm because it depicts monolayer adsorption (Figure 3.6a). A type I isotherm is obtained when $P/P_o < 1$ and $c > 1$ in the BET equation, where P/P_o is the partial pressure value and c is the BET constant, which is related to the adsorption energy of the first monolayer and varies from solid to solid. The characterization of microporous materials, those with pore diameters less than 2 nm, gives this type of isotherm.

Type II isotherm

A type II isotherm (Figure 3.6b) is very different than the Langmuir model. The flatter region in the middle represents the formation of a monolayer. A type II isotherm is obtained when $c > 1$ in the BET equation. This is the most common isotherm obtained when using the BET technique. At very low pressures, the micropores fill with nitrogen gas. At the knee, monolayer formation is beginning, and multilayer formation occurs at medium pressure. At the higher pressures, capillary condensation occurs.

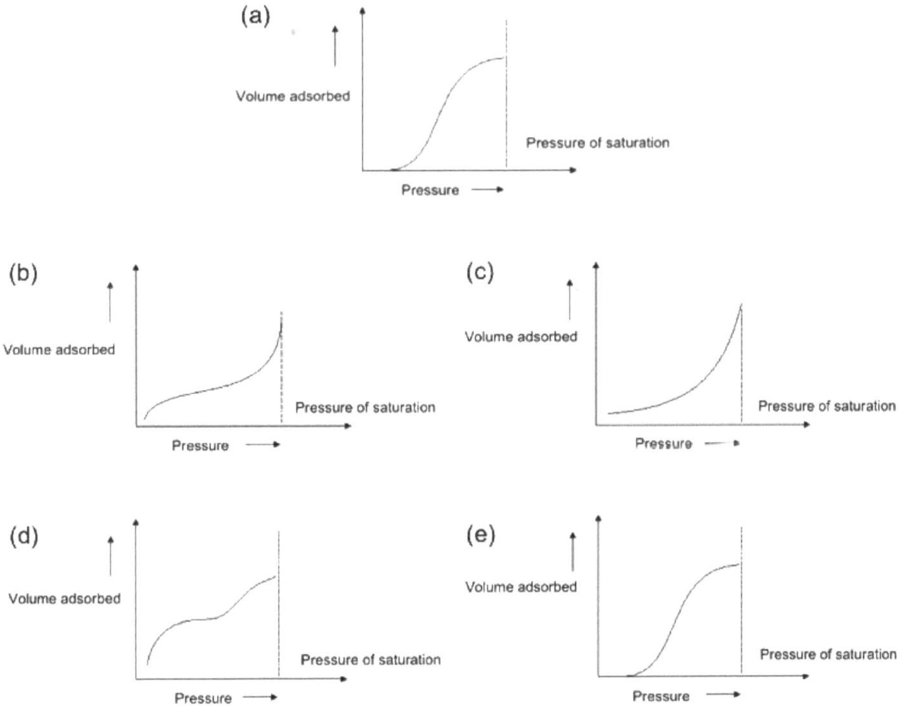

Figure 3.6: The isotherm plots for the volume of gas adsorbed onto the surface of the sample as pressure increases: (a) Type I, (b) Type II, (c) Type III, (d) Type IV, and (e) Type V. Adapted from S. Brunauer, L. S. Deming, W. E. Deming, and E. Teller, On a theory of the van der Waals adsorption of gases. *J. Am. Chem. Soc.***, 1940, 62, 1723. Copyright: American Chemical Society (1940).**

Type III isotherm

A type III isotherm (Figure 3.6c) is obtained when the $c < 1$ and shows the formation of a multilayer. Because there is no asymptote in the curve, no monolayer is formed, and BET is not applicable.

Type IV isotherm

Type IV isotherms (Figure 3.6d) occur when capillary condensation occurs. Gases condense in the tiny capillary pores of the solid at pressures below the saturation pressure of the gas. At the lower pressure regions, it shows the formation of a monolayer followed by a formation of multilayers. BET surface area characterization of mesoporous materials, which are materials with pore diameters between 2 - 50 nm, gives this type of isotherm.

Type V isotherm

Type V isotherms (Figure 3.6e) are very similar to type IV isotherms and are not applicable to BET.

Calculations

The BET equation,

$$\frac{1}{X[(P_0/P)-1]]} = \frac{1}{X_mC} + \frac{C-1}{X_mC}(P/P_0)$$

uses the information from the isotherm to determine the surface area of the sample, where X is the weight of nitrogen adsorbed at a given relative pressure (P/Po), X_m is monolayer capacity, which is the volume of gas adsorbed at standard temperature and pressure (STP), and C is constant. STP is defined as 273 K and 1 atm.

Multi-point BET

Ideally five data points, with a minimum of three data points, in the P/P_0 range 0.025 to 0.30 should be used to successfully determine the surface area using the BET equation. At relative pressures higher than 0.5, there is the onset of capillary condensation, and at relative pressures that are too low, only monolayer formation is occurring. When the BET equation is plotted, the graph should be of linear with a positive slope. If such a graph is not obtained, then the BET method was insufficient in obtaining the surface area. The slope and y-intercept can be obtained using least squares regression.

The monolayer capacity X_m can be calculated with,

$$X_m = \frac{1}{s+i} = \frac{C-1}{C_s}$$

Once X_m is determined, the total surface area S_t can be calculated with the following equation, where L_{av} is Avogadro's number, A_m is the cross sectional area of the adsorbate and equals 0.162 nm^2 for an absorbed nitrogen molecule, and M_v is the molar volume and equals 22414 mL,

$$S = \frac{X_m L_{av} A_m}{M_v}$$

Single point BET can also be used by setting the intercept to 0 and ignoring the value of C. The data point at the relative pressure of 0.3 will match up the best with a multipoint BET. Single point BET can be used over the more accurate multipoint BET to determine the appropriate relative pressure range for multi-point BET.

Sample preparation and experimental setup

Prior to any measurement the sample must be degassed to remove water and other contaminants before the surface area can be accurately measured. Samples are degassed in a vacuum at high temperatures. The highest temperature possible that will not damage the sample's structure is usually chosen in order to shorten the degassing time. IUPAC recommends that samples be degassed for at least 16 hours to ensure that unwanted vapors and gases are removed from the surface of the sample. Generally, samples that can withstand higher temperatures without structural changes have smaller degassing times. A minimum of 0.5 g of sample is required for the BET to successfully determine the surface area.

Samples are placed in glass cells to be degassed and analyzed by the BET machine. Glass rods are placed within the cell to minimize the dead space in the cell. Sample cells typically come in sizes of 6, 9 and 12 mm and come in different shapes. 6 mm cells are usually used for ne powders, 9 mm cells for larger particles and small pellets and 12 mm are used for large pieces that cannot be further reduced. The cells are placed into heating mantles and connected to the outgas port of the machine.

After the sample is degassed, the cell is moved to the analysis port (Figure 3.7). Dewars of liquid nitrogen are used to cool the sample and maintain it at a constant temperature. A low temperature must be maintained so that the interaction between the gas molecules and the surface of the sample will be strong enough for measurable amounts of adsorption to occur. The adsorbate, nitrogen gas in this case, is injected into the sample cell with a calibrated piston. The dead volume in the sample cell must be calibrated before and after each measurement. To do that, helium gas is used for a blank run, because helium does not adsorb onto the sample.

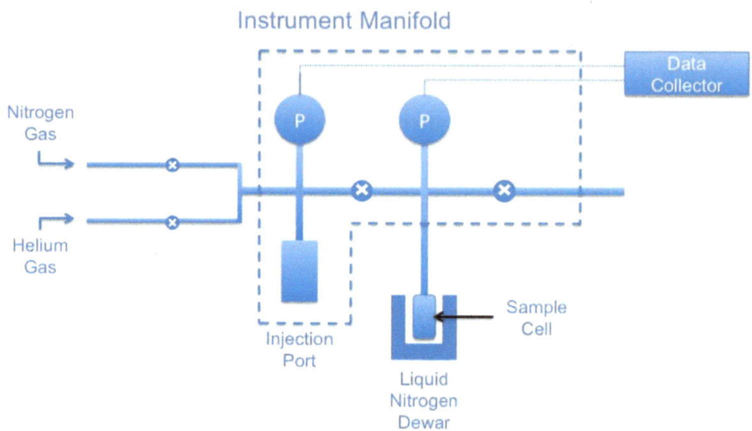

Figure 3.7: Schematic representation of the BET instrument. The degasser is not shown.

Shortcomings of BET

The BET technique has some disadvantages when compared to NMR, which can also be used to measure the surface area of nanoparticles. BET measurements can only be used to determine the surface area of dry powders. This technique requires a lot of time for the adsorption of gas molecules to occur. A lot of manual preparation is required.

The surface area determination of metal-organic frameworks

The BET technique was used to determine the surface areas of metal-organic frameworks (MOFs), which are crystalline compounds of metal ions coordinated to organic molecules. Possible applications of MOFs, which are porous, include gas purification and catalysis. An isoreticular MOF (IRMOF) with the chemical formula $Zn_4O(pyrene-1,2-dicarboxylate)_3$ (Figure 3.8) was used as an example to see if BET could accurately determine the surface area of microporous materials. The predicted surface area was calculated directly from the geometry of the crystals and agreed with the data obtained from the BET isotherms. Data was collected at a constant temperature of 77 K and a type II isotherm (Figure 3.6b) was obtained. The isotherm data obtained from partial pressure range of 0.05 to 0.3 is plugged into the BET equation to obtain the BET plot (Figure 3.9).

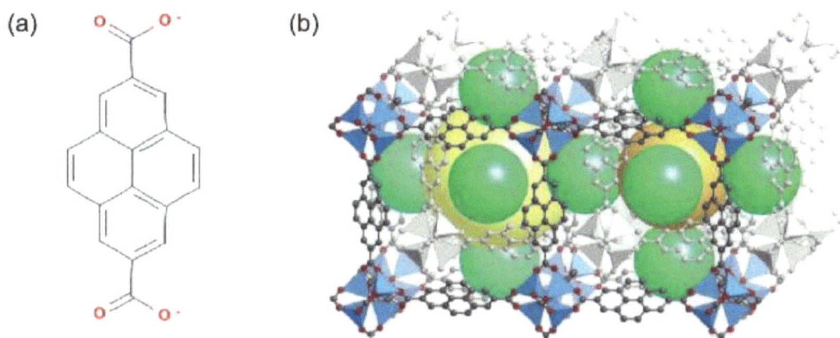

Figure 3.8: The structure of (a) the pyrene-1,2-dicarboxylate linkage unit and (b) the catenated IRMOF-13 Zn_4O(pyrene-1,2-dicarboxylate)$_3$. Orange and yellow represent non-catenated pore volumes. Green represents catenated pore volume. Adapted from Y. S. Bae, R. Q. Snurr, and O. Yazaydin, Evaluation of the BET method for determining surface areas of MOFs and zeolites that contain ultra-micropores. *Langmuir*, 2010, 26, 5479. Copyright: American Chemical Society (2010).

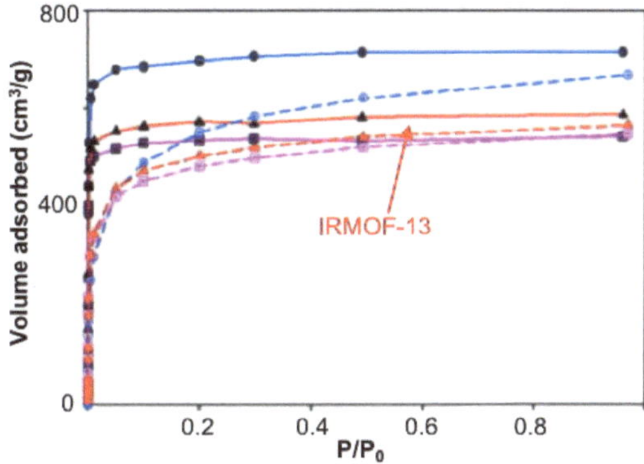

Figure 3.9: The BET isotherms of the zeolites and metal-organic frameworks. IRMOF-13 is symbolized by the black triangle and red line. Adapted from Y. S. Bae, R. Q. Snurr, and O. Yazaydin, Evaluation of the BET method for determining surface areas of MOFs and zeolites that contain ultra-micropores. *Langmuir*, 2010, 26, 5479. Copyright: American Chemical Society (2010).

The monolayer capacity is determined to be 391.2 cm^3/g. Now that Xm is known, then the surface area determined is 1702.3 m^2/g.

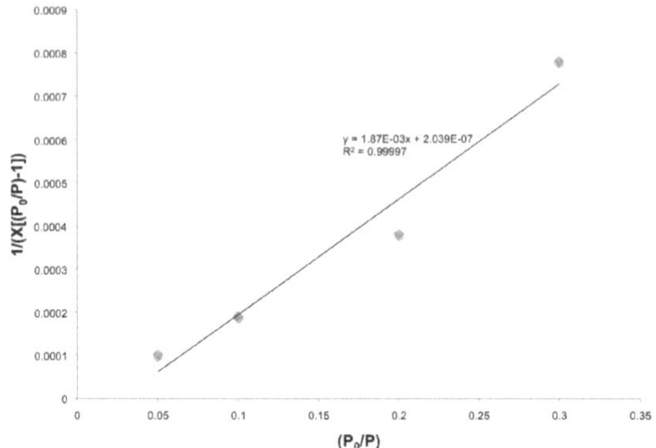

Figure 3.14: BET plot of IRMOF-13 using points collected at the pressure range 0.05 to 0.3. The equation of the best-fit line and R^2 value are shown. Data from Y. S. Bae, R. Q. Snurr, and O. Yazaydin, Evaluation of the BET method for determining surface areas of MOFs and zeolites that contain ultra-micropores. *Langmuir*, 2010, 26, 5479.

Bibliography

Y. S. Bae, R. Q. Snurr, and O. Yazaydin, Evaluation of the BET method for determining surface areas of MOFs and zeolites that contain ultra-micropores. *Langmuir*, 2010, **26**, 5479.

S. Brunauer, L. S. Deming, W. E. Deming, and E. Teller, On a theory of the van der Waals adsorption of gases. *J. Am. Chem. Soc.*, 1940, **62**, 1723.

S. Brunauer, P. H. Emmett, and E. Teller, Adsorption of gases in multimolecular layers. *J. Am. Chem. Soc.*, 1938, **60**, 309.

B. H. Davis, Paul H. Emmett (1900-1985): six decades of catalysis. *J. Phys. Chem.*, 1986, **90**, 4701.

C. D. Jones and A. R. Barron, Porosity, crystal phase, and morphology of nanoparticle derived alumina as a function of the nanoparticle's carboxylate substituent. *Mater. Chem. Phys.*, 2007, **104**, 460.

K. S. Sing, Obituary: Stephen Brunauer, (1903-1986). *Langmuir*, 1987, **3**, 2.

Chapter 4: Zeta Potential Analysis

Gaowei Chen, Pavan M. V. Raja and Andrew R. Barron

Introduction

The physical properties of colloids (nanoparticles) and suspensions are strongly dependent on the nature and extent of the particle-liquid interface. The behavior of aqueous dispersions between particles and liquid is especially sensitive to the ionic and electrical structure of the interface.

Zeta potential is a parameter that measures the electrochemical equilibrium at the particle-liquid interface. It measures the magnitude of electrostatic repulsion/attraction between particles and thus, it has become one of the fundamental parameters known to affect stability of colloidal particles. It should be noted that that term stability, when applied to colloidal dispersions, generally means the resistance to change of the dispersion with time. Figure 4.1 illustrates the basic concept of zeta potential.

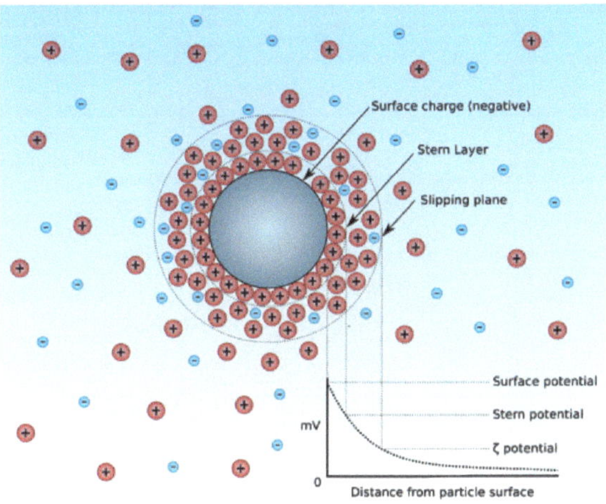

Figure 4.1: Schematic representation of the ionic concentration and potential difference as a function of distance from the charged surface of a particle suspended in a dispersion medium.

From the fundamental theory's perspective, zeta potential is the electrical potential in the interfacial double layer (DL) at the location of the slipping plane (shown in Figure 4.1). We can regard zeta potential as the potential difference between the dispersion medium and the stationary layer of the fluid attached to the particle layer. Therefore, in experimental concerns, zeta potential is key factor in processes such as the preparation of colloidal dispersions, utilization of colloidal phenomena and the destruction of unwanted colloidal dispersions. Moreover, zeta potential analysis and measurements nowadays have a lot of real-world applications. In the field of biomedical research, zeta potential measurement, in contrast to chemical methods of analysis which can disrupt the organism, has the particular merit of providing information referring to the outermost regions of an organism. It is also largely utilized in water purification and treatment. Zeta potential analysis has established optimum coagulation conditions for removal of particulate matter and organic dyestuffs from aqueous waste products.

History and development

Zeta potential is a scientific term for electrokinetic potential in colloidal dispersions. In prior literature, it is usually denoted using the Greek letter zeta, Z, hence it has obtained the name zeta potential as Z-potential. The earliest theory for calculating Zeta potential from experimental data was developed by Marian Smoluchowski in 1903 (Figure 4.2). Even till today, this theory is still the most well-known and widely used method for calculating zeta potential.

Figure 4.2: Portrait of Polish physicist Marian Smoluchowski (1872 - 1917) pioneer of statistical physics.

Interestingly, this theory was originally developed for electrophoresis. Later on, people started to apply his theory in calculation of zeta potential. The main reason that this theory is powerful is because of its universality and validity for dispersed particles of any shape and any concentration. However, there still some limitations to this early theory as it was mainly determined experimentally. The main limitations are that Smoluchowski's theory neglects the contribution of surface conductivity and only works for particles which have sizes much larger than the interface layer, denoted as κ_a ($1/\kappa$ is the Debye length and a is the particle radius).

Overbeek (Figure 4.3) and Booth as early pioneers in this direction started to develop more theoretical and rigorous electrokinetic theories that were able to incorporate surface conductivity for electrokinetic applications. Modern rigorous electrokinetic theories that are valid for almost any κ_a mostly have since been developed.

Figure 4.3: Dutch physical chemist Jan Theodoor Gerard Overbeek (1911 - 2007).

Principle of zeta potential analysis

Electrokinetic phenomena

Because an electric double-layer (EDL) exists between a surface and solution, then any relative motion between the rigid and mobile parts of the EDL will result in the generation of an electrokinetic potential. As described above, zeta potential is essentially an electrokinetic potential which rises from electrokinetic phenomena. So, it is important to understand different situations where electrokinetic potential can be produced. There are generally four fundamental ways which zeta potential can be produced, via electrophoresis, electroosmosis, streaming potential, and sedimentation potential as shown from Figure 4.4.

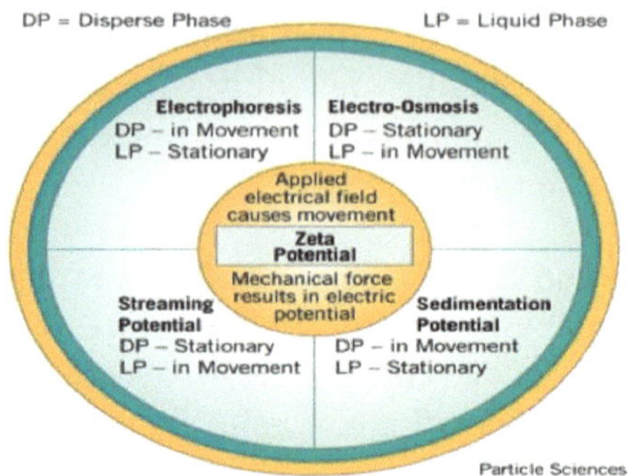

Figure 4.4: Relationship between the four types of electrokinetic phenomena. Reproduced from D. Fairhurst, An overview of the zeta potential - Part 2: measurement. *Am. Pharm. Rev.*, **2013. Copyright: CompareNetworks (2013).**

Calculations of zeta potential

There are many different ways of calculating zeta potential. In this section, the methods of calculating zeta potential in electrophoresis and electroosmosis will be introduced.

Zeta potential in electrophoresis

Electrophoresis is the movement of charged colloidal particles or polyelectrolytes, immersed in a liquid, under the influence of an external electric field. In such case, the electrophoretic velocity, v_e (ms^{-1}) is the velocity during electrophoresis and the electrophoretic mobility, u_e (m^{2}V^{-1}s^{-1}) is the magnitude of the velocity divided by the magnitude of the electric field strength. The mobility is counted positive if the particles move toward lower potential and negative in the opposite case. And therefore, we have the relationship $v_e = u_e E$, where E is the externally applied field.

Thus, the formula accounted for zeta potential in electrophoresis case is given in,

$$u_e = \frac{\varepsilon_{rs}\varepsilon_0\zeta}{\eta}$$

where ε_{rs} is the relative permittivity of the electrolyte solution, ε_0 is the electric permittivity of vacuum and η is the viscosity.

$$v_e = \frac{\varepsilon_{rs}\varepsilon_0\zeta}{\eta} E$$

There are two cases regarding the size of κ_a:

- $\kappa_a < 1$: the formula is similar,

$$u_e = (2/3)\,\frac{\varepsilon_{rs}\varepsilon_0\zeta}{\eta}$$

- $\kappa_a > 1$: the formula is rather complicated, and we need to solve equation for zeta potential, where $y^{e\zeta} = e\zeta/kT$, m is about 0.15 for aqueous solution.

$$\frac{3}{2}\frac{\eta e}{\varepsilon_{rs}\varepsilon_0\,kT}u_e = \frac{3}{2}y^{ek} - \frac{6\left[\frac{y^{ek}}{2} - \frac{\ln 2}{\zeta}\{1 - \exp\,(-\zeta y^{ek})\}\right]}{2 + \frac{ka}{1+3m/\zeta^2}\exp\,\left(\frac{-\zeta y^{ek}}{2}\right)}$$

Electroosmosis is the motion of a liquid through an immobilized set of particles, a porous plug, a capillary, or a membrane, in response to an applied electric field. Similar to electrophoresis, it has the electroosmotic velocity, v_{eo} (ms^{-1}) as the uniform velocity of the liquid far from the charged interface. Usually, the measured quantity is the volume flow rate of liquid divided by electric field strength, Q_{eo},E (m^{3}V^{-1}s^{-1}) or divided by the electric current, Q_{eo},I (m^{3}C^{-1}). Therefore, the relationship is given by,

$$Q_{eo} = \iint v_{eo}\,dS$$

Thus, the formula accounted for Zeta potential in electroosmosis is given in,

$$Q_{eo,E} = \frac{-\varepsilon_{rs}\varepsilon_0 \zeta}{\eta} Ac$$

As with electrophoresis there are two cases regarding the size of κ_a:
- $\kappa_a \gg 1$ and there is no surface conduction, where Ac is the cross-section area and K_L is the bulk conductivity of particle.
- $\kappa_a < 1$ where $\Delta u = K^\sigma/K_L$ is the Dukhin number account for surface conductivity, K^σ is the surface conductivity of the particle.

$$Q_{eo,I} = \frac{-\varepsilon_{rs}\varepsilon_0 \zeta}{\eta} \frac{1}{K_L}$$

$$Q_{eo,I} = \frac{-\varepsilon_{rs}\varepsilon_0 \zeta}{\eta} \frac{1}{K_L \, (1 + 2\Delta u)}$$

Relationship between zeta potential and particle stability in electrophoresis

Using the above theoretical methods, we can calculate zeta potential for particles in electrophoresis. The following table summarizes the stability behavior of the colloid particles with respect to zeta potential. Thus, we can use zeta potential to predict the stability of colloidal particles in the electrokinetic phenomena of electrophoresis.

Zeta potential (mV)	Stability behavior of the particles
0 to ±5	Rapid coagulation or occlusion
±10 to ±30	Incipient instability
±30 to ±40	Moderate stability
±40 to ±60	Good stability
More than ±61	Excellent stability

Table 4.1: Stability behavior of the colloid particles with respect to zeta potential.

Instrumentation

Figure 4.5 shows example of typical commercial zeta potential analyzer for electrophoresis.

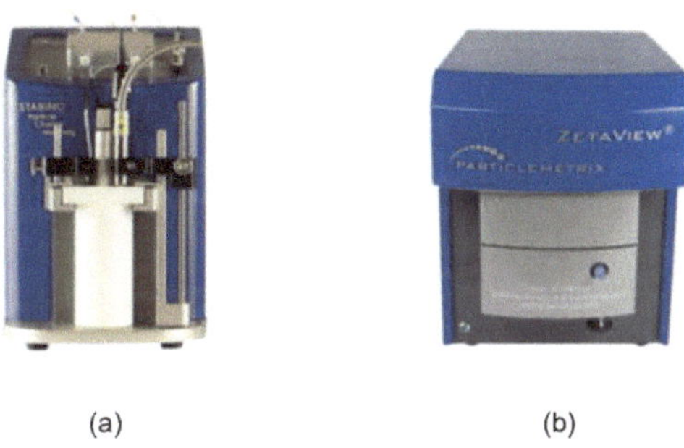

(a)   (b)

Figure 4.5: Typical zeta potential analyzer for electrophoresis: (a) Stabin® and (b) ZetaView®.

The inside measuring principle is described in the following diagram, which shows the detailed mechanism of zeta potential analyzer (Figure 4.6).

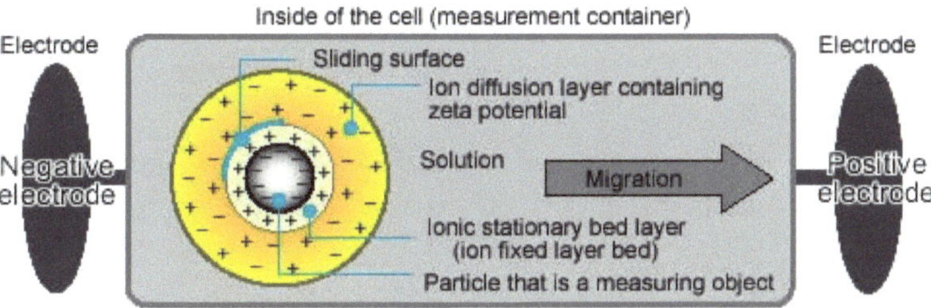

Figure 4.6: Mechanism of zeta potential analyzer for electrophoresis (Microtec Co., Ltd.).

When a voltage is applied to the solution in which particles are dispersed, particles are attracted to the electrode of the opposite polarity, accompanied by the fixed layer and part of the di use double layer, or internal side of the "sliding surface". Using the following formula below of this specific Analyzer and the computer program, we can obtain the zeta potential for electrophoresis using,

$$\zeta = (4\pi\eta/\varepsilon) \times U \times 300 \times 300 \times 1000$$

where ζ is the zeta potential (mV), η is the viscosity of the solution, ε is the dielectric constant, U is the electrophoretic mobility, as defined by,

$$\zeta = (4\pi\eta/\varepsilon) \times U \times 300 \times 300 \times 1000$$

where v is the speed of the particle (cm/sec), V is the voltage (V), and L is the length of the electrode.

$$U = vL/V$$

Bibliography

F. Booth, Theory of Electrokinetic Effects. *Nature*, 1948, **161**, 83.

A. V. Delgado, F. Gonzalez-Caballero, R. J. Hunter, L. K. Koopal and J. Lyklema, Measurement and interpretation of electrokinetic phenomena. *J. Colloid Interface Sci.*, 2007, **309**, 194.

D. Fairhurst, An overview of the zeta potential - Part 2: measurement. *Am. Pharm. Rev.*, 2013.

D. Fairhurst and V. Ribitsch, Zeta potential measurements of irregular shape solid materials. *ACS Symposium Series*, 1991, **472**, 337.

R. J. Hunter, *Zeta Potential in Colloid Science: Principles and Applications*, School of Chemistry, University of Sydney, Sydney, Australia.

A. Morfesis, M. A. Jacobson, R. Frollini, M. Helgeson, J. Billica and R. K. Gertig, Role of zeta (ζ) potential in the optimization of water treatment facility operations. *Ind. Eng. Chem. Res.*, 2009, **48**, 2305.

M. Predota, L. M. Machesky and J. D. Wesolowski, Molecular origins of the zeta potential. *Langmuir*, 2016, **32**, 10189.

M. Smoluchowski, Contribution to the theory of electro-osmosis and related phenomena. *Bull. Int. Acad. Sci.*, 1903, **3**, 184.

E. J. W. Verwey and J. Th. *G. Overbeek*, Elsevier (1948).

Chapter 5: Dynamic Light Scattering

Yilun Li and Andrew R. Barron

Introduction

Dynamic light scattering (DLS), which is also known as photon correlation spectroscopy (PCS) or quasi-elastic light scattering (QLS), is a spectroscopy method used in the fields of chemistry, biochemistry, and physics to determine the size distribution of particles (polymers, proteins, colloids, etc.) in solution or suspension. In the DLS experiment, normally a laser provides the monochromatic incident light, which impinges onto a solution with small particles in Brownian motion (named after Robert Brown, Figure 5.1).

Figure 5.1: Scottish botanist and palaeobotanist Robert Brown (1773 - 1858) who first described the phenomenon later named Brownian motion.

Through the Rayleigh scattering process (Figure 5.2), particles whose sizes are sufficiently small compared to the wavelength of the incident light will diffract the incident light in all direction with different wavelengths and intensities as a function of time. Since the scattering pattern of the light is highly correlated to the size distribution of the analyzed particles, the size-related information of the sample could be then acquired by mathematically processing the spectral characteristics of the scattered light.

Figure 5.2: British scientist John William Strutt, 3rd Baron Rayleigh (1842 - 1919) who received the Nobel Prize in Physics in 1904 "for his investigations of the densities of the most important gases and for his discovery of argon".

Theory

The theory of DLS can be introduced utilizing a model system of spherical particles in solution. According to the Rayleigh scattering (Figure 5.3), when a sample of particles with diameter smaller than the wavelength of the incident light, each particle will diffract the incident light in all directions, while the intensity I is determined by,

$$I = I_0 \frac{1 + \cos^2\theta}{2R^2} (\frac{2\pi}{\lambda})^4 (\frac{n^2 - 1}{n^2 + 2})^2 r^6$$

where I_0 and λ is the intensity and wavelength of the unpolarized incident light, R is the distance to the particle, θ is the scattering angel, n is the refractive index of the particle, and r is the radius of the particle.

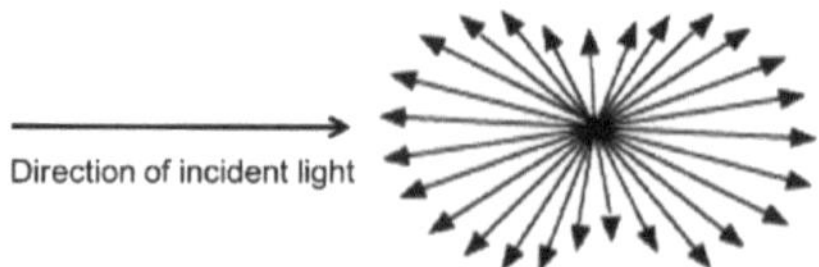

Figure 5.3: Scheme of Rayleigh scattering.

If that diffracted light is projected as an image onto a screen, it will generate a speckle" pattern (Figure 5.4); the dark areas represent regions where the diffracted light from the particles arrives out of phase interfering destructively and the bright area represent regions where the diffracted light arrives in phase interfering constructively.

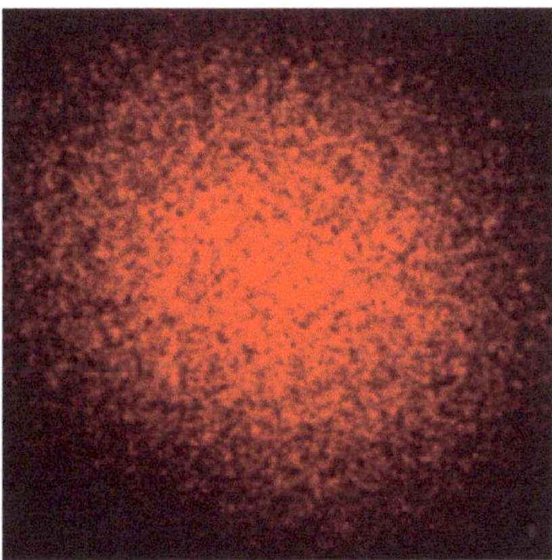

Figure 5.4: Typical speckle pattern.

In practice, particle samples are normally not stationary but moving randomly due to collisions with solvent molecules as described by the Brownian motion,

$$\overline{(\Delta x)^2} = 2Dt$$

where (Δx) is the mean squared displacement in time t, and D is the diffusion constant, which is related to the hydrodynamic radius a of the particle according to the Stokes-Einstein equation,

$$D = \frac{k_B T}{6\pi\mu a}$$

where kB is Boltzmann constant, T is the temperature, and μ is viscosity of the solution. Importantly, for a system undergoing Brownian motion, small particles should di use faster than large ones.

As a result of the Brownian motion, the distance between particles is constantly changing and this results in a Doppler shift between the frequency of the incident light and the frequency of the scattered light. Since the distance between particles also affects the phase overlap/interfering of the diffracted light, the brightness and darkness of the spots in the speckle pattern will in turn fluctuate in intensity as a function of time when the particles change position with respect to each other. Then, as the rate of these intensity fluctuations depends on how fast the particles are moving (smaller particles di use faster), information about the size distribution of particles in the solution could be acquired by processing the fluctuations of the intensity of scattered light. Figure 5.5 shows the hypothetical fluctuation of scattering intensity of larger particles and smaller particles.

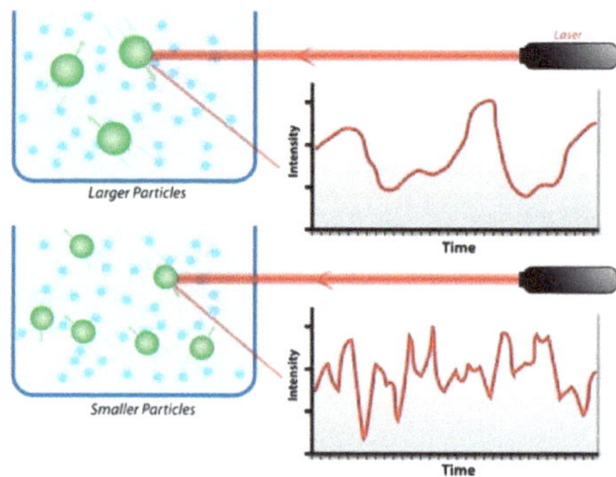

Figure 5.5: Hypothetical fluctuation of scattering intensity of larger particles and smaller particles.

In order to mathematically process the fluctuation of intensity, there are several principles/terms to be understood. First, the intensity correlation function is used to describe the rate of change in scattering intensity by comparing the intensity I(t) at time t to the intensity I(t + τ) at a later time (t + τ), and is quantified and normalized by,

$$G_2(\tau) = \langle I(t)I(t + \tau) \rangle$$

and

$$g_2(\tau) = \frac{\langle I(t)I(t + \tau) \rangle}{\langle I(t) \rangle^2}$$

where braces indicate averaging over t.

Second, since it is not possible to know how each particle moves from the fluctuation, the electric field correlation function is instead used to correlate the motion of the particles relative to each other, and is defined by,

$$G_1(\tau)= \langle E(t)E(t+\tau)\rangle$$

and,

$$g_1(\tau)= \frac{\langle E(t)E(t+\tau)\rangle}{\langle E(t)E(t)\rangle}$$

where E(t) and E(t + τ) are the scattered electric fields at times t and t + τ.

For a monodisperse system undergoing Brownian motion, $g_1(\tau)$ will decay exponentially with a decay rate Γ which is related by Brownian motion to the diffusivity by,

$$g_1(\tau)= e^{-\Gamma\tau}$$

$$\Gamma = -Dq^2$$

$$q = \frac{4\pi n}{\lambda}\sin\frac{\theta}{2}$$

where q is the magnitude of the scattering wave vector and q^2 reflects the distance the particle travels, n is the refraction index of the solution, and θ is angle at which the detector is located.

For a polydisperse system however, $g_1(\tau)$ can no longer be represented as a single exponential decay and must be represented as an intensity-weighed integral over a distribution of decay rates G(Γ) by,

$$g_1(\tau)= \int_0^\infty G(\Gamma)e^{-\Gamma\tau}d\Gamma$$

where G(Γ) is normalized,

$$\int_0^\infty G(\Gamma)d\Gamma = 1$$

Third, the two correlation functions above can be equated using the Seigert relationship based on the principles of Gaussian random processes (which the scattering light usually is), and can be expressed as,

$$g_2(\tau) = B + \beta[g_1(\tau)]^2$$

where β is a factor that depends on the experimental geometry, and B is the long-time value of $g_2(\tau)$, which is referred to as the baseline and is normally equal to 1. Figure 5.6 shows the decay of $g_2(\tau)$ for small size sample and large size sample.

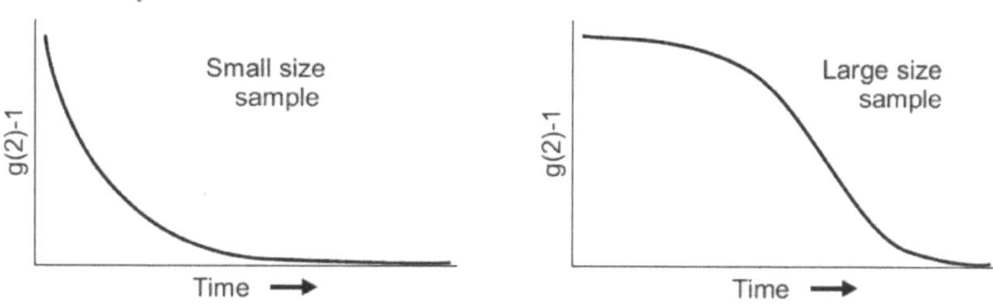

Figure 5.6: Decay of $g_2(\tau)$ for small size sample and large size sample. Malvern Instruments Ltd., Zetasizer Nano Series User Manual, 2004. Copyright: Malvern Instruments Ltd. (2004).

When determining the size of particles in solution using DLS, $g_1(\tau)$ is calculated based on the time- dependent scattering intensity and is converted through the Seigert relationship to $g_1(\tau)$ which usually is an exponential decay or a sum of exponential decays. The decay rate Γ is then mathematically determined from the $g_1(\tau)$ curve, and the value of diffusion constant D and hydrodynamic radius a can be easily calculated afterwards.

Experimental

Instrument of DLS

In a typical DLS experiment, light from a laser passes through a polarizer to de ne the polarization of the incident beam and then shines on the scattering medium. When the sizes of the analyzed particles are sufficiently small compared to the wavelength of the incident light, the incident light will scatter in all directions known as the Rayleigh scattering. The scattered light then passes through an analyzer, which selects a given polarization and finally enters a detector, where the position of the detector defines the scattering angle θ. In addition, the intersection of the incident beam and the beam intercepted by the detector defines a scattering region of volume V. As for the detector used in these experiments, a phototube is normally used whose dc output is proportional to the intensity of the scattered light beam. Figure 5.7 shows a schematic representation of the light-scattering experiment.

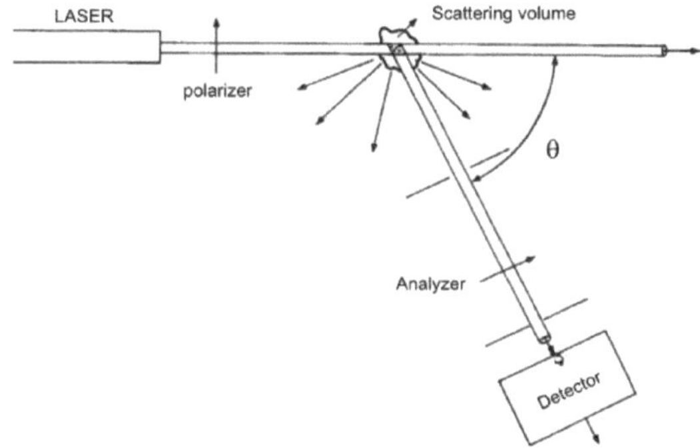

Figure 5.7: A schematic representation of the light-scattering experiment. Adapted from B. J. Berne and R. Pecora, *Dynamic Light Scattering: With Applications to Chemistry, Biology, and Physics*, Dover, Mineola, NY (2000). Copyright: Dover Publications (2000).

In modern DLS experiments, the scattered light spectral distribution is also measured. In these cases, a photomultiplier is the main detector, but the pre- and post-photomultiplier systems differ depending on the frequency change of the scattered light. The three different methods used are filter (f > 1 MHz), homodyne (f > 10 GHz), and heterodyne methods (f < 1 MHz), as schematically illustrated in Figure 5.8. Note that that homodyne and heterodyne methods use no monochromator of filter between the scattering cell and the photomultiplier, and optical mixing techniques are used for heterodyne method. shows the schematic illustration of the various techniques used in light-scattering experiments.

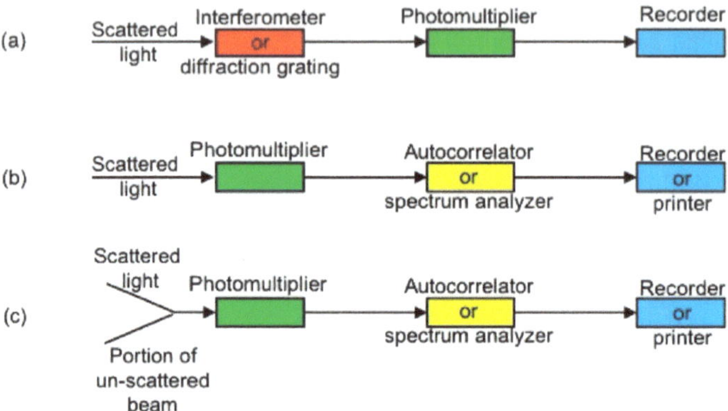

Figure 5.8: Schematic illustration of the various techniques used in light-scattering experiments: (a) filter methods; (b) homodyne; (c) heterodyne. Adapted from B. J. Berne and R. Pecora, *Dynamic Light Scattering: With Applications to Chemistry, Biology, and Physics*, Dover, Mineola, NY (2000). Copyright: Dover Publications (2000).

As for an actual DLS instrument, take the Zetasizer Nano (Malvern Instruments Ltd.) as an example (Figure 5.9), it actually looks like nothing other than a big box, with components of power supply, optical unit (light source and detector), computer connection, sample holder, and accessories.

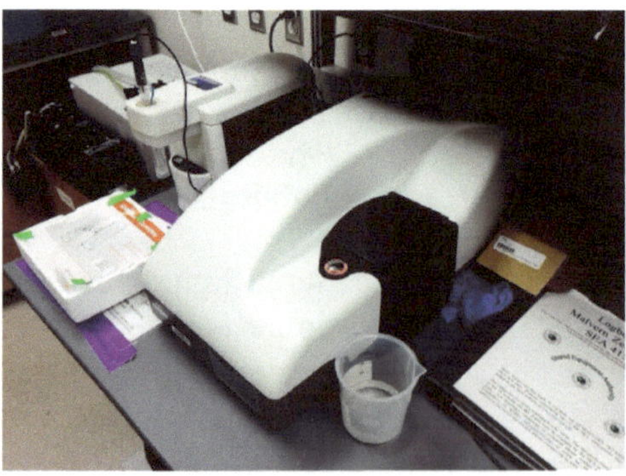

Figure 5.9: Photo of a DLS instrument at Rice University (Zetasizer Nano, Malvern Instruments Ltd.).

Sample preparation

Although different DLS instruments may have different analysis ranges, we are usually looking at particles with a size range of nm to μm in solution. For several kinds of samples, DLS can give results with rather high confidence, such as monodisperse suspensions of unaggregated nanoparticles that have radius > 20 nm, or polydisperse nanoparticle solutions or stable solutions of aggregated nanoparticles that have radius in the 100 - 300 nm range with a polydispersity index of 0.3 or below. For other more challenging samples such as solutions containing large aggregates, bimodal solutions, very dilute samples, very small nanoparticles, heterogeneous samples, or unknown samples, the results given by DLS could not be really reliable, and one must be aware of the strengths and weaknesses of this analytical technique.

Then, for the sample preparation procedure, one important question is how much materials should be submitted, or what is the optimal concentration of the solution. Generally, when doing the DLS measurement, it is important to submit enough amount of material in order to obtain sufficient signal, but if the sample is overly concentrated, then light scattered by one particle might be again scattered by another (known as multiple scattering), and make the data processing less accurate. An ideal sample submission for DLS analysis has a volume of 1 2 mL and is sufficiently concentrated as to have strong color hues, or opaqueness/turbidity in the case of a white or black sample. Alternatively, 100 - 200 μL of highly concentrated sample can be diluted to 1 mL or analyzed in a low-volume microcuvette.

In order to get high quality DLS data, there are also other issues to be concerned with. First is to minimize particulate contaminants, as it is common for a single particle contaminant to scatter a million times more than a suspended nanoparticle, by using ultra high purity water or solvents, extensively rinsing pipettes and containers, and sealing sample tightly. Second is to filter the sample through a 0.2 or 0.45 μm filter to get away of the visible particulates within the sample solution. Third is to avoid probe sonication to prevent the particulates ejected from the sonication tip and use the bath sonication instead.

Measurement

Now that the sample is readily prepared and put into the sample holder of the instrument, the next step is to actually do the DLS measurement. Generally, the DLS instrument will be provided with software that can help you to do the measurement rather easily, but it is still worthwhile to understand the important parameters used during the measurement.

Firstly, the laser light source with an appropriate wavelength should be selected. As for the Zetasizer Nano series (Malvern Instruments Ltd.), either a 633 nm red laser or a 532 nm green laser is available. One should keep in mind that the 633 nm laser is least suitable for blue samples, while the 532 nm laser is least suitable for red samples, since otherwise the sample will just absorb a large portion of the incident light.

Then, for the measurement itself, one has to select the appropriate stabilization time and the duration time. Normally, longer striation/duration time can result in more stable signal with less noises, but the time cost should also be considered. Another important parameter is the temperature of the sample, as many DLS instruments are equipped with the temperature-controllable sample holders, one can actually measure the size distribution of the data at different temperatures and obtain extra information about the thermal stability of the sample analyzed.

Next, as is used in the calculation of particle size from the light scattering data, the viscosity and refraction index of the solution are also needed. Normally, for solutions with low concentration, the viscosity and refraction index of the solvent/water could be used as an approximation.

Finally, to get data with better reliability, the DLS measurement on the same sample will normally be conducted multiple times, which can help eliminate unexpected results and also provide additional error bar of the size distribution data.

Data analysis

Although size distribution data could be readily acquired from the software of the DLS instrument, it is still worthwhile to know about the details about the data analysis process.

Cumulant method

The decay rate Γ is mathematically determined from the $g_1(\tau)$ curve; if the sample solution is monodispersed, $g_1(\tau)$ could be regard as a single exponential decay function $e^{-\Gamma\tau}$, and the decay rate Γ can be in turn easily calculated. However, in most of the practical cases, the sample solution is always polydispersed, $g_1(\tau)$ will be the sum of many single exponential decay functions with different decay rates, and then it becomes significantly difficult to conduct the fitting process.

There is, however, a few methods developed to meet this mathematic challenge: linear fit and cumulant expansion for mono-modal distribution, exponential sampling and CONTIN regularization for non-monomodal distribution. Among all these approaches, cumulant expansion is most common method and will be illustrated in detail in this section.

Generally, the cumulant expansion method is based on two relations: one between $g1(\tau)$ and the moment- generating function of the distribution, and one between the logarithm of $g1(\tau)$ and the cumulant-generating function of the distribution.

To start with, the form of $g1(\tau)$ is equivalent to the definition of the moment-generating function $M(-\tau,\Gamma)$ of the distribution $G(\Gamma)$,

$$g_1(\tau)= \int_0^\infty G(\Gamma)e^{-\Gamma\tau}d\Gamma = M(-\tau,\Gamma)$$

The mth moment of the distribution $mm(\Gamma)$ is given by the mth derivative of $M(-\tau,\Gamma)$ with respect to τ,

$$m_m(\Gamma)= \int_0^\infty G(\Gamma)\Gamma^m e^{-\Gamma\tau}d\Gamma\big|_{-\tau=0}$$

Similarly, the logarithm of $g1(\tau)$ is equivalent to the definition of the cumulant-generating function $K(-\tau,\Gamma)$, EQ, and the mth cumulant of the distribution $km(\Gamma)$ is given by the mth derivative of $K(-\tau,\Gamma)$ with respect to τ,

$$\ln g_1(\tau)= \ln M(-\tau,\Gamma)= K(-\tau,\Gamma)$$

$$k_m(\Gamma)= \frac{d^m K(-\tau,\Gamma)}{d(-\tau)^m}\big|_{-\tau=0}$$

By making use of that the cumulants, except for the first, are invariant under a change of origin, the $k_m(\Gamma)$ could be rewritten in terms of the moments about the mean as,

$$k_1(\tau)= \int_0^\infty G(\Gamma)\Gamma d\Gamma = \bar{\Gamma}$$

$$k_2(\tau) = \mu_2$$

$$k_3(\tau) = \mu_3$$

where μm are the moments about the mean, defined as given in,

$$\mu_m = \int_0^\infty G(\Gamma)(\Gamma - \bar{\Gamma})^m d\Gamma$$

Based on the Taylor expansion of $K(-\tau,\Gamma)$ about $\tau = 0$, the logarithm of $g1(\tau)$ is given as,

$$\ln g_1(\tau)= K(-\tau,\Gamma)= -\bar{\Gamma}\tau + \frac{k_2}{2!}\tau^2 - \frac{k_3}{3!}\tau^3 + \frac{k_4}{4!}\tau^4 \cdots$$

Importantly, if look back at the Seigert relationship in the logarithmic form,

$$\ln(g_2(\tau)\text{-}B) = \ln\beta + 2\ln g_1(\tau)$$

The measured data of $g_2(\tau)$ could be fitted with the parameters of k_m using the relationship of,

$$\ln(g_2(\tau)-B)= \ln\beta + 2(-\bar{\Gamma}\tau + \frac{k_2}{2!}\tau^2 - \frac{k_3}{3!}\tau^3 \cdots)$$

where Γ, (k_1), k_2, and k_3 describes the average, variance, and skewness (or asymmetry) of the decay rates of the distribution, and polydispersity index $\gamma = (k_2/\Gamma)$ is used to indicate the width of the distribution, and parameters beyond k_3 are seldom used to prevent over fitting the data. Finally, the size distribution can be easily calculated from the decay rate distribution as described above. Figure 5.10 shows an example of data fitting using the cumulant method for vesicles formed with [(2R)-3-hexadecanoyloxy-2-[(Z)-octadec-9-enoyl]oxypropyl] 2-(trimethylazaniumyl)ethyl phosphate (POPC, Figure 5.11).

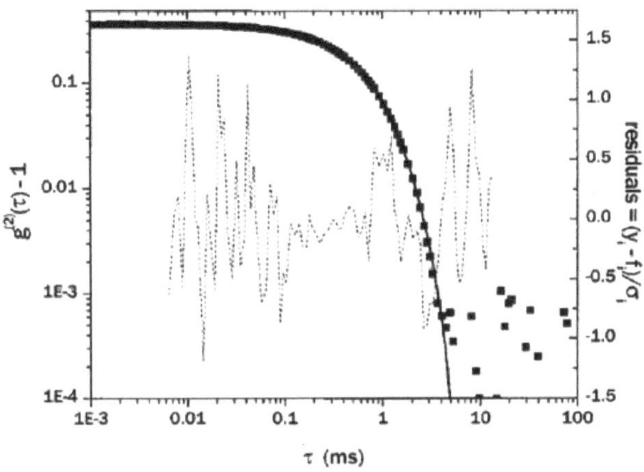

Figure 5.10: Sample data taken for POPC (Figure 5.11) vesicles formed by extrusion through polycarbonate mem- branes. The dashed curve shows the weighted residuals: the difference of the fit from the data divided by the uncertainty in each point. Adapted from B. J. Frisken, Revisiting the method of cumulants for the analysis of dynamic light-scattering data. *Appl. Optics*, 2001, 40, 4087. Copyright: Optical Society of America (2001).

Figure 5.11: Structure of [(2R)-3-hexadecanoyloxy-2-[(Z)-octadec-9-enoyl]oxypropyl] 2-(trimethylazaniumyl)ethyl phosphate (POPC).

When using the cumulant expansion method however, one should keep in mind that it is only suitable for monomodal distributions (Gaussian-like distribution centered about the mean), and for non-monomodal distributions, other methods like exponential sampling and CONTIN regularization should be applied in-stead.

Three index of size distribution

Now that the size distribution is able to be acquired from the fluctuation data of the scattered light using cumulant expansion or other methods, it is worthwhile to understand the three kinds of distribution index usually used in size analysis: number weighted distribution, volume weighted distribution, and intensity weighted distribution.

First of all, based on all the theories discussed above, it should be clear that the size distribution given by DLS experiments is the intensity weighted distribution, as it is always the intensity of the scattering that is being analyzed. So, for intensity weighted distribution, the contribution of each particle is related to the intensity of light scattered by that particle. For example, using Rayleigh approximation, the relative contribution for very small particles will be proportional to a^6.

For number weighted distribution, given by image analysis as an example, each particle is given equal weighting irrespective of its size, which means proportional to a^0. This index is most useful where the absolute number of particles is important, or where high resolution (particle by particle) is required.

For volume weighted distribution, given by laser diffraction as an example, the contribution of each particle is related to the volume of that particle, which is proportional to a^3. This is often extremely useful from a commercial perspective as the distribution represents the composition of the sample in terms of its volume/mass, and therefore its potential money value.

When comparing particle size data for the same sample represented using different distribution index, it is important to know that the results could be very different from number weighted distribution to intensity weighted distribution. This is clearly illustrated in the example below (Figure 5.12), for a sample consisting of equal numbers of particles with diameters of 5 nm and 50 nm. The number weighted distribution gives equal weighting to both types of particles, emphasizing the presence of the finer 5 nm particles, whereas the intensity weighted distribution has a signal one million times higher for coarse 50 nm particles. The volume weighted distribution is intermediate between the two.

Furthermore, based on the different orders of correlation between the particle contribution and the particle size a, it is possible to convert particle size data from one type of distribution to another type of distribution, and that is also why the DLS software can also give size distributions in three different forms (number, volume, and intensity), where the first two kinds are actually deducted from the raw data of intensity weighted distribution.

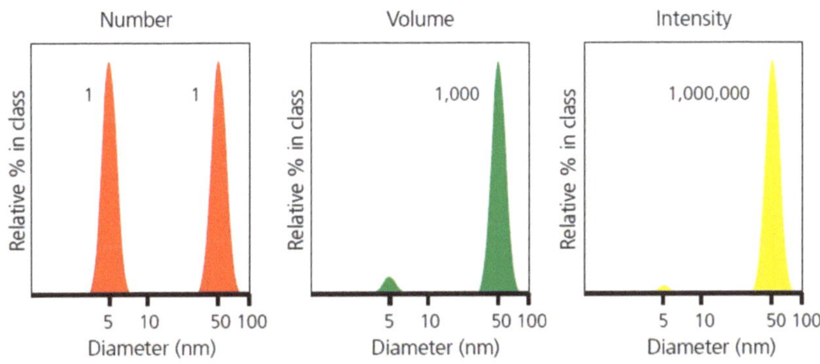

Figure 5.12: Example of number, volume and intensity weighted particle size distributions for the same sample. Malvern Instruments Ltd., A Basic Guide to Particle Characterization, 2012. Copyright: Malvern Instrument Ltd. (2012).

Comparison with TEM and AFM

Since DLS is not the only method available to determine the size distribution of particles, it is also necessary to compare DLS with the other common-used general sizing techniques, especially TEM and AFM.

First of all, it has to be made clear that both TEM and AFM measure particles that are deposited on a substrate (Cu grid for TEM, mica for AFM), while DLS measures particles that are dispersed in a solution. In this way, DLS will be measuring the bulk phase properties and give a more comprehensive information about the size distribution of the sample. For AFM or TEM, it is very common that a relatively small sampling area is analyzed, and the size distribution on the sampling area may not be the same as the size distribution of the original sample depending on how the particles are deposited.

On the other hand, for DLS, the calculating process is highly dependent on the mathematical and physical assumptions and models, which is, monomodal distribution (cumulant method) and spherical shape for the particles, the results could be inaccurate when analyzing non-monomodal distributions or non- spherical particles. Yet, since the size determining process for AFM or TEM is nothing more than measuring the size from the image and then using the statistic, these two methods can provide much more reliable data when dealing with irregular samples.

Another important issue to consider is the time cost and complication of size measurement. Generally speaking, the DLS measurement should be a much easier technique, which requires less operation time and also cheaper equipment, and it could be really troublesome to analysis the size distribution data coming out from TEM or AFM images without specially programmed software.

In addition, there are some special issues to consider when choosing size analysis techniques. For example, if the originally sample is already on a substrate (synthesized by the CVD method), or the particles could not be stably dispersed within solution, apparently the DLS method is not suitable. Also, when the particles tend to have a similar imaging contrast against the substrate (carbon nanomaterials on TEM grid) or tend to self-assemble and aggregate on the surface of the substrate, the DLS approach might be a better choice.

In general research work, the best way to obtain size distribution analysis is to combine these analyzing methods and get complimentary information from different aspects. One thing to keep in mind, since the DLS actually measures the hydrodynamic radius of the particles, the size from DLS measurement is always larger than the size from AFM or TEM measurement. As a conclusion, the comparison between DLS and AFM/TEM is shown in Table 5.1.

	DLS	**AFM/TEM**
Sample preparation	Solution	Substrate
Measurement	Easy	Difficult
Sampling	Bulk	Small area
Shape of particles	Sphere	No requirement
Polydispersity	Low	No requirement
Size range	nm to μm	nm to μm
Size information	Hydrodynamic radius	Physical size

Table 5.1: Comparison between DLS, AFM and TEM.

Bibliography

B. J. Berne and R. Pecora, *Dynamic Light Scattering: With Applications to Chemistry, Biology, and Physics*; John Wiley & Sons, Inc., New York (2000).

R. F. Domingos, M. A. Baalousha, Y. Ju-Nam, M. M. Reid, N. Tufenkji, J. R. Lead, G. G. Leppard, and K. J. Wilkinson, Characterizing manufactured nanoparticles in the environment: multimethod determination of particle sizes. *Environ. Sci. Technol.*, 2009, **43**, 7277.

B. J. Frisken, Revisiting the method of cumulants for the analysis of dynamic light-scattering data. *Appl. Optics*, 2001, **40**, 4087.

C. M. Hoo, N. Starostin, P. West, and M. L. Mecartney, A comparison of atomic force microscopy (AFM) and dynamic light scattering (DLS) methods to characterize nanoparticle size distributions. *J. Nanopart. Res.*, 2008, **10**, 89.

D. E. Koppel, Analysis of macromolecular polydispersity in intensity correlation spectroscopy: the method of cumulant. *J. Chem. Phys.*, 1972, **57**, 4814.

Malvern Instruments Ltd., *Zetasizer Nano Series User Manual*, 2004.

Malvern Instruments Ltd., *A Basic Guide to Particle Characterization*, 2012.

C. T. Vogelson and A. R. Barron, Particle size control and dependence on solution pH of carboxylate-alumoxane nanoparticles. *J. Non-Cryst. Solids*, 2001, **290**, 216.

Z. Wang, B. Tan, I. Hussain, N. Schaeffer, M. F. Wyatt, M. Brust, and A. I. Cooper, Design of polymeric stabilizers for size-controlled synthesis of monodisperse gold nanoparticles in water. *Langmuir*, 2006, **23**, 885.

Chapter 6: Viscosity

Desmond Schipper and Andrew R. Barron

Introduction

All liquids have a natural internal resistance to flow termed viscosity. Viscosity is the result of frictional interactions within a given liquid and is commonly expressed in two different ways.

Dynamic viscosity

The first is dynamic viscosity, also known as absolute viscosity, which measures a fluid's resistance to flow. In precise terms, dynamic viscosity is the tangential force per unit area necessary to move one plane past another at unit velocity at unit distance apart. As one plane moves past another in a fluid, a velocity gradient is established between the two layers (Figure 6.1). Viscosity can be thought of as a drag coefficient proportional to this gradient.

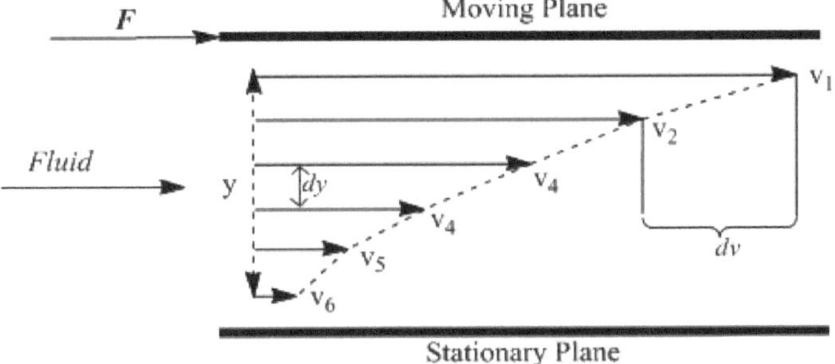

Figure 6.1: Fluid dynamics as one plane moves relative to a stationary plane through a liquid. The moving plane has area A and requires force F to overcome the fluid's internal resistance.

The force necessary to move a plane of area A past another in a fluid is given by,

$$F = \eta A(V/Y)$$

where V is the velocity of the liquid, Y is the separation between planes, and η is the dynamic viscosity.

V/Y also represents the velocity gradient (sometimes referred to as shear rate). Force over area is equal to τ, the shear stress, so the equation simplifies to,

$$\tau = \eta(V/Y)$$

For situations where V does not vary linearly with the separation between plates, the differential formula based on Newton's equations is given in,

$$\tau = \eta(\delta V/\delta Y)$$

Kinematic viscosity

Kinematic viscosity, the other type of viscosity, requires knowledge of the density, ρ, and is given by,

$$\nu = \eta/\rho$$

where ν is the kinematic viscosity and η is the dynamic viscosity.

Viscosity is commonly expressed in Stokes, Poise, Saybolt Universal Seconds, degree Engler, and SI units.

Units of viscosity

Dynamic viscosity

The SI units for dynamic (absolute) viscosity is given in units of $N \cdot S/m^2$, $Pa \cdot S$, or $kg/(m \cdot s)$, where N stands for Newton and Pa for Pascal. Poise are metric units expressed as $dyne \cdot s/cm^2$ or $g/(m \cdot s)$. They are related to the SI unit by $g/(m \cdot s) = 1/10 \ Pa \cdot S$. 100 centipoise, the centipoise (cP) being the most used unit of viscosity, is equal to one Poise.

Table 6.1 shows the interconversion factors for dynamic viscosity. Table 6.2 lists the dynamic viscosities of several liquids at various temperatures in centipoise. The effect of the temperature on viscosity is clearly evidenced in the drastic drop in viscosity of water as the temperature is increased from near ambient to 60 °C. Ketchup has a viscosity of 1000 cP at 30 °C or more than 1000 times that of water at the same temperature!

Unit	Pa·S	dyne·s/cm² or g/(m·s) (Poise)	Centipoise (cP)
Pa·S	1	10	1000
dyne·s/cm² or g/(m·s) (Poise)	0.1	1	100
Centipoise (cP)	0.001	0.01	1

Table 6.1: The interconversion factors for dynamic viscosity.

Liquid	η (cP)	Temperature (°C)
Water	0.89	25
Water	0.47	60
Milk	2.0	18
Olive oil	107.5	20
Toothpaste	70,000-100,000	18
Ketchup	1000	30
Custard	1,500	85-90
Crude oil (WTI)[a]	7	15

Table 6.2: Viscosities of common liquids. [a]At 0% evaporation volume.

Kinematic viscosity

The CGS unit for kinematic viscosity is the Stoke which is equal to 10^{-4} M^2/s. Dividing by 100 yields the more commonly used centistoke. The SI unit for viscosity is m^2/s. The Saybolt Universal second is commonly used in the oil field for petroleum products represents the time required to fleflux 60 mL from a Saybolt Universal viscometer at a fixed temperature according to ASTM D-88. The Engler scale is often used in Britain and quantifies the viscosity of a given liquid in comparison to water in an Engler viscometer for 200 cm^3 of each liquid at a set temperature.

Newtonian versus non-Newtonian fluids

One of the invaluable applications of the determination of viscosity is identifying a given liquid as Newtonian or non-Newtonian in nature.
- Newtonian liquids are those whose viscosities remain constant for all values of applied shear stress.
- Non-Newtonian liquids are those liquids whose viscosities vary with applied shear stress and/or time.

Moreover, non-Newtonian liquids can be further subdivided into classes by their viscous behavior with shear stress:

- Pseudoplastic fluids whose viscosity decreases with increasing shear rate,
- Dilatants in which the viscosity increases with shear rate,
- Bingham plastic fluids, which require some force threshold be surpassed to begin to flow and which thereafter flow proportionally to increasing shear stress.

Measuring viscosity

Viscometers are used to measure viscosity. There are seven different classes of viscometer:

- Capillary viscometers.
- Orifice viscometers.
- High temperature high shear rate viscometers.
- Rotational viscometers.
- Falling ball viscometers.
- Vibrational viscometers.
- Ultrasonic Viscometers.

Capillary viscometers

Capillary viscometers are the most widely used viscometers when working with Newtonian fluids and measure the flow rate through a narrow, usually glass tube. In some capillary viscometers, an external force is required to move the liquid through the capillary; in this case, the pressure difference across the length of the capillary is used to obtain the viscosity coefficient.

Capillary viscometers require a liquid reservoir, a capillary of known dimensions, a pressure controller, a flow meter, and a thermostat be present. These viscometers include: Modified Ostwald viscometers, Suspended-level viscometers, and Reverse-flow viscometers and measure kinematic viscosity.

The equation governing this type of viscometry is the Pouisille law,

$$Q = \frac{\pi \Delta P a^4}{8 \eta l}$$

where Q is the overall flowrate, ΔP, the pressure difference, a, the internal radius of the tube, η, the dynamic viscosity, and l the path length of the fluid.

Here, Q is equal to V/t; the volume of the liquid measured over the course of the experiment divided by the time required for it to move through the capillary where V is volume and t is time.

For gravity-type capillary viscometers, those relying on gravity to move the liquid through the tube rather than an applied force,

$$\eta = \frac{\pi g h a^4}{8lV} \rho t$$

is used to find viscosity, obtained by substituting the relation with the experimental values,

$$\Delta P = \rho g h$$

where P is pressure, ρ is density, g is the gravitational constant, and h is the height of the column.

An example of a capillary viscometer (Ostwald viscometer) is shown in Figure 6.2.

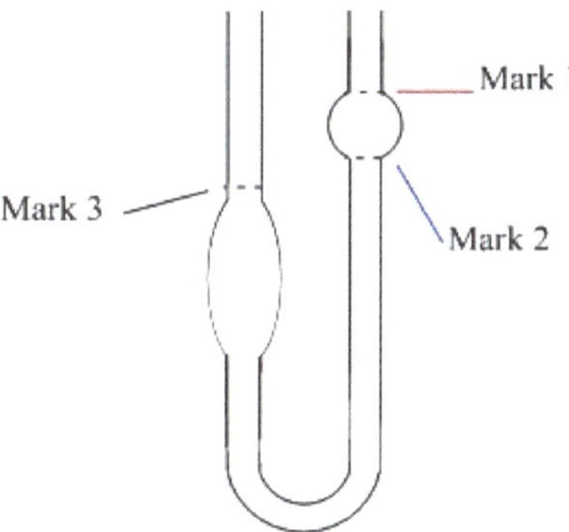

Figure 6.2: The capillary, submerged in an isothermal bath, is filled until the liquid lies at Mark 3. The liquid is then drawn up through the opposite side of the tube. The time it takes for the liquid to travel from Mark 2 to Mark 1 is used to compute the viscosity.

Orifice viscometers

Commonly found in the oil industry, orifice viscometers consist of a reservoir, an orifice, and a receiver. These viscometers report viscosity in units of efflux time as the measurement consists of measuring the time it takes for a given liquid to travel from the orifice to the receiver. These instruments are not accurate as the set-up does not ensure that the pressure on the liquid remains constant and there is energy lost to friction at the orifice. The most common types of these viscometer include Redwood, Engler, Saybolt, and Ford cup viscometers. A Saybolt viscometer is represented in Figure 6.3.

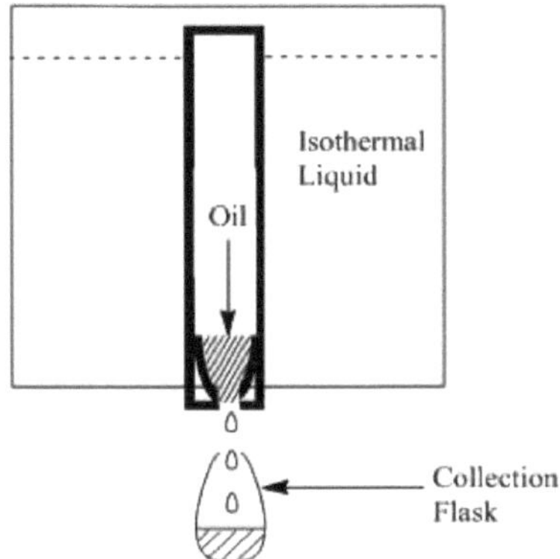

Figure 6.3: The time it takes for a 60 mL collection ask to ll is used to determine the viscosity in Saybolt units.

High temperature, high shear rate viscometers

These viscometers, also known as cylinder-piston type viscometers are employed when viscosities above 1000 poise, need to be determined, especially of non-Newtonian fluids. In a typical set-up, fluid in a cylindrical reservoir is displaced by a piston. As the pressure varies, this type of viscometry is well-suited for determining the viscosities over varying shear rates, ideal for characterizing fluids whose primary environment is a high temperature, high shear rate environment, e.g., motor oil. A typical cylinder-piston type viscometer is shown in Figure 6.4.

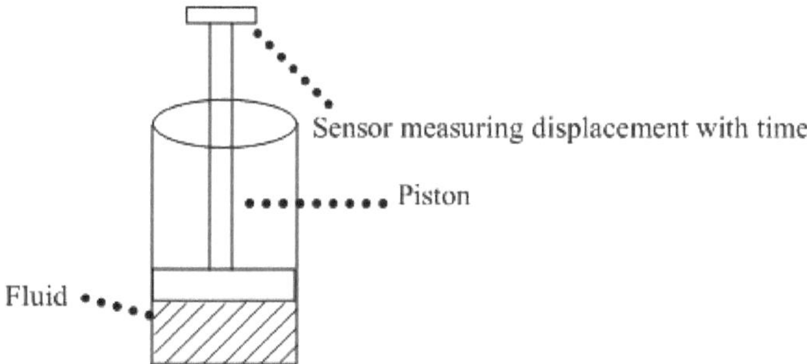

Figure 6.4: A typical cylinder-piston type viscometer.

Rotational viscometers

Well-suited for non-Newtonian fluids, rotational viscometers measure the rate at which a solid rotates in a viscous medium. Since the rate of rotation is controlled, the amount of force necessary to spin the solid can be used to calculate the viscosity. They are advantageous in that a wide range of shear stresses and temperatures and be sampled across. Common rotational viscometers include: the coaxial-cylinder viscometer, cone and plate viscometer, and coni-cylinder viscometer. A cone and plate viscometer are shown in Figure 6.5.

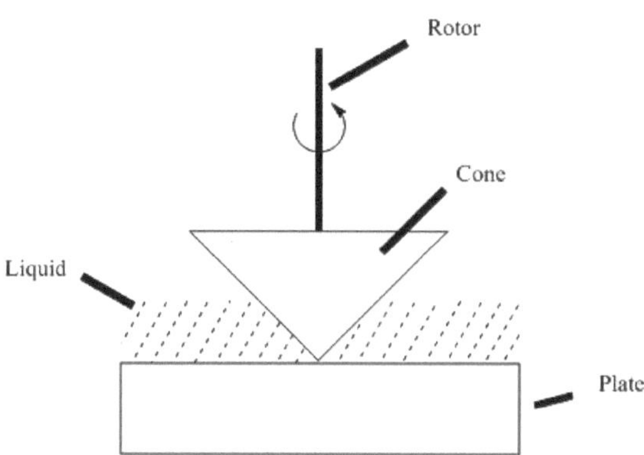

Figure 6.5: A cone is spun by a rotor in a liquid paste along a plate. The response of the rotation of the cone is measured, thereby determining viscosity.

Falling ball viscometer

This type of viscometer relies on the terminal velocity achieved by a balling falling through the viscous liquid whose viscosity is being measured. A sphere is the simplest object to be used because its velocity can be determined by rearranging Stokes' law,

$$6\pi r \eta \upsilon = (4/3)\pi r^3(\sigma-\rho)g$$

to

$$\eta = \frac{(4/3)\pi r^2(\sigma-\rho)g}{6\pi v}$$

where r is the sphere's radius, η the dynamic viscosity, v the terminal velocity of the sphere, σ the density of the sphere, ρ the density of the liquid, and g the gravitational constant

A typical falling ball viscometric apparatus is shown in Figure 6.6.

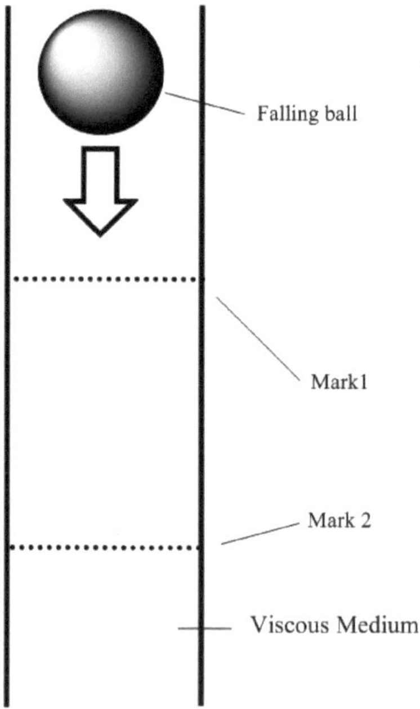

Figure 6.6: The time taken for the falling ball to pass from mark 1 to mark 2 is used to obtain viscosity measurements.

Vibrational viscometers

Often used in industry, these viscometers are attached to fluid production processes where a constant viscosity quality of the product is desired. Viscosity is measured by the damping of an electrochemical resonator immersed in the liquid to be tested. The resonator is either a cantilever, oscillating beam, or a tuning fork. The power needed to keep the oscillator oscillating at a given frequency, the decay time after stopping the oscillation, or by observing the difference when waveforms are varied are respective ways in which this type of viscometer works. A typical vibrational viscometer is shown in Figure 6.7.

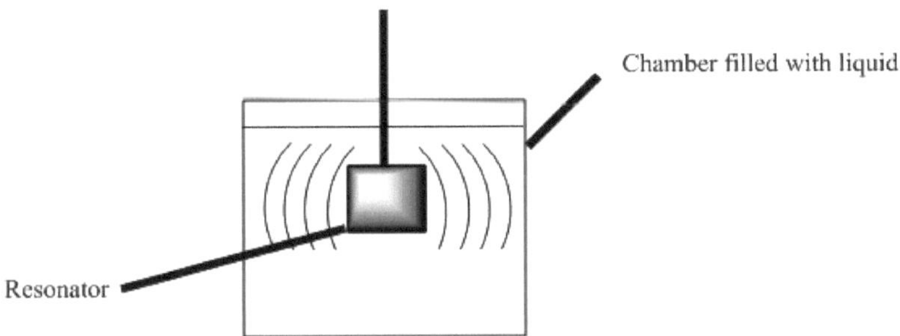

Figure 6.7: A resonator produces vibrations in the liquid whose viscosity is to be tested. An external sensor detects the vibrations with time, characterizing the material's viscosity in real time.

Ultrasonic viscometers

This type of viscometer is most like vibrational viscometers in that it is obtaining viscosity information by exposing a liquid to an oscillating system. These measurements are continuous and instantaneous. Both ultrasonic and vibrational viscometers are commonly found on liquid production lines and constantly monitor the viscosity.

Bibliography

D. S. Viswanath, T. K. Gosh, D. H. L. Prasad, N. V. K. Dutt, K. Y. Rani. *Viscosity of Liquids: Theory, Estimation, Experiment, and Data*, Springer, 1st edn., 2007.

C. W. Macosko, *Rheology: Principles, Measurements, and Applications*, Wiley-VCH, New Jersey, 1st edn., 1994.

F. A. Morrison, *Understanding Rheology*, Oxford University Press, New York, 1st edn., 2001.

Spring Handbook for Experimental Fluid Mechanics, Ed. C. Tropea, A.L. Yarin, J.F. Foss, Springer, 1st edn., 2007.

Chapter 7: Electrochemistry

Xianyu Li, Kendahl L. Walz Mitra, Anjli Kumar, Pavan M. V. Raja
and Andrew R. Barron

Fundamentals of electrochemistry

A chemical reaction that involves a change in the charge of a chemical species is called an *electrochemical reaction*. As the name suggests, these reactions involve electron transfer between chemicals. Many of these reactions occur spontaneously when the various chemicals come in contact with one another. In order to force a nonspontaneous electrochemical reaction to occur, a driving force needs to be provided. This is because every chemical species has a relative reduction potential. These values provide information on the ability of the chemical to take extra electrons. Conversely, we can think if relative oxidation potentials, which indicate the ability of a chemical to give away electrons.

It is important to note that values for the oxidation or reduction potentials are relative to, and need to be defined against, a reference reaction. The common standard is the hydrogen electrode (SHE) (Figure 7.1), which undergoes the reaction,

$$2H^+_{(aq)} + 2\ e^- \rightarrow H_2$$

Although this reaction is shown as a reduction, the SHE can act as either the anode or the cathode, depending on the relative oxidation/reduction potential of the other electrode/electrolyte combination. The term standard in SHE requires a supply of hydrogen gas bubbled through the electrolyte at a pressure of 1 atm and an acidic electrolyte with H^+ activity equal to 1 (usually assumed to be $[H^+]$ = 1 mol/liter), see Figure 7.1.

A list of standard reduction potentials for common electrochemical half-reactions is given in Table 7.1. Nonspontaneous electrochemical systems, often called electrolytic cells, as mentioned previously, require a driving force to occur. This driving force is an applied voltage, which forces reduction of the chemical that is less likely to gain an electron.

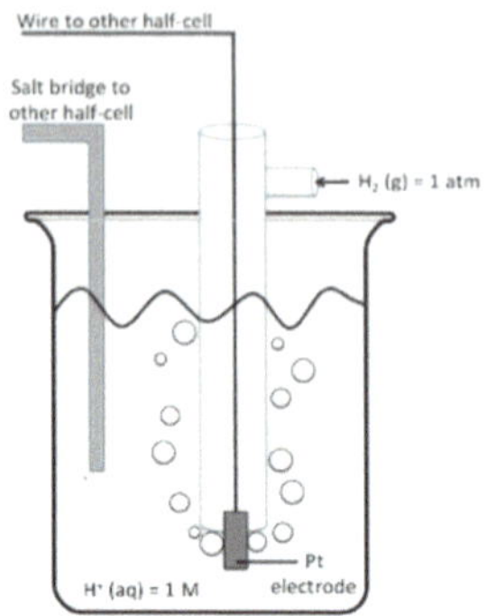

Figure 7.1: A schematic diagram of a standard hydrogen electrode.

Oxidant	Reductant	E^0 (V versus SHE)
$2H_2O + 2e^-$	H_2 (g) $+2OH^-$	-0.8227
Cu_2O (s) $+ H_2O + 2e^-$	$2Cu$ (s) $+ 2OH^-$	-0.360
$Sn^{4+} + 2e^-$	Sn^{2+}	+0.15
$Cu^{2+} + 2e^-$	Cu (s)	+0.337
O_2 (g) $+2H^+ + 2e^-$	H_2O_2 (aq)	+0.70

Table 7.1: List of standard reduction potentials of various half reactions.

Design of an electrochemical cell

A schematic of an electrochemical cell is seen in Figure 7.2. In its simplest form an electrochemical cell must have at least two electrodes:

- a cathode, where the reduction half-reaction takes place,
- an anode, where the oxidation half-reaction occurs.

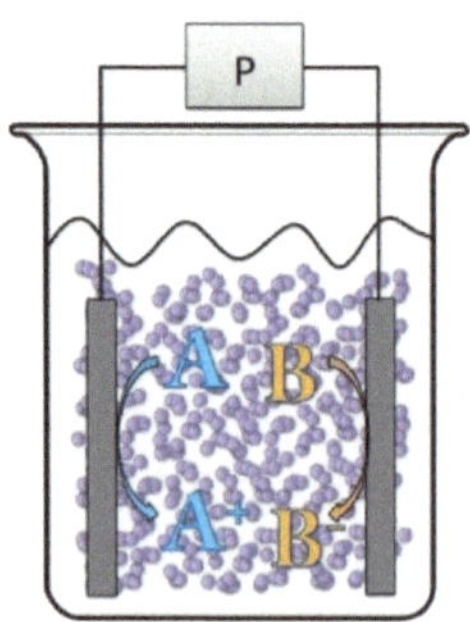

Figure 7.2: Schematic of an electrochemical cell.

Examples of half reactions can be seen in Table 7.1. The two electrodes are electrically connected in two ways:

- the electrolyte solution,
- the external wire.

The electrolyte solution typically includes a small amount of the electroactive analyte (the chemical species that will actually participate in electron transfer) and a large amount of supporting electrolyte (the chemical species that assist in the movement of charge but are not actually involved in electron transfer).

The external wire provides a path for the electrons to travel from the oxidation half-reaction to the reduction half-reaction. As mentioned previously, when an electrolytic reaction (nonspontaneous) is being forced to occur a voltage needs to be applied. This requires the wires to be connected to a potentiostat. As its name suggests, a potentiostat controls voltage (i.e., *potentio* = potential measured in volts). The components of an electrochemical cell and their functions are also given in Table 7.2.

Component	Function
Electrode	Interface between ions and electrons
Anode	Electrode at which the oxidation half reaction takes place
Cathode	Electrode at which the reduction half reaction takes place
Electrolyte solution	Solution that contains supporting electrolyte and electroactive analyte
Supporting electrolyte	Not a part of the faradaic process; only a part of the capacitive process
Electroactive analyte	The chemical species responsible for all faradaic current
Potentiostat	DC voltage source; sets the potential difference between the cathode and the anode
Wire	Connects the electrodes to the potentiostat

Table 7.2: Various components of an electrochemical cell and their respective functions.

Cyclic voltammetry

Overview

Cyclic voltammetry is a key electrochemical technique that, among its other uses, can be employed to examine the kinetics of oxidation-reduction reactions in electrochemical systems. Specifically, data collected with cyclic voltammetry can be used to determine the rate of reaction. In its simplest form, this technique requires a simple three electrode cell and a potentiostat (Figure 7.3).

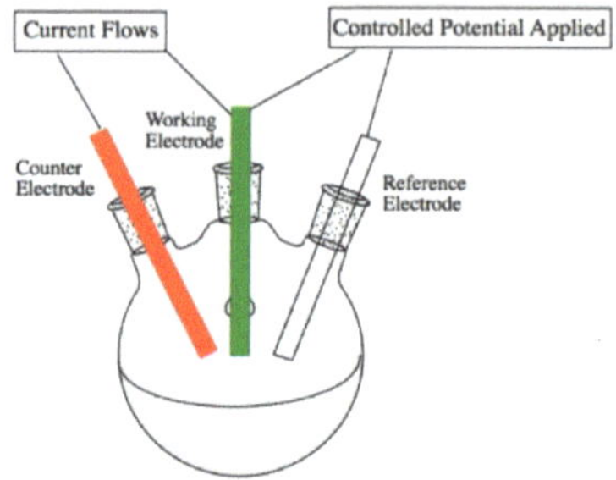

Figure 7.3: A simple three electrode cell.

A potential applied to the working electrode is varied linearly with time and the response in the current is measured. Typically, the potential is cycled between two values once in the forward direction and once in the reverse direction. For example, in Figure 7.4a, the potential is cycled between V_1 and V_2 with the forward scan moving from positive to negative potential and the reverse scan moving from negative to positive potential.

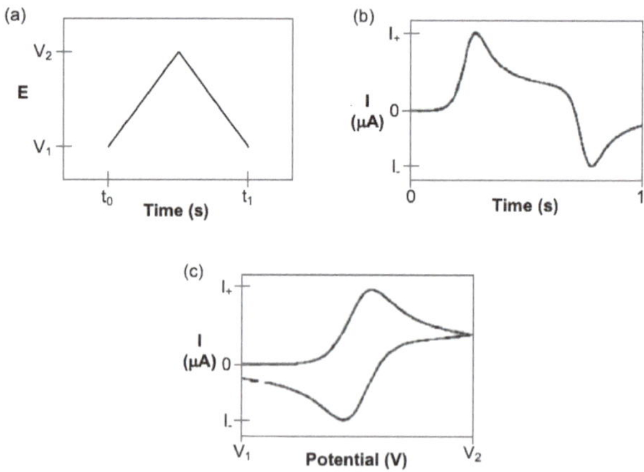

Figure 7.4: Plots of (a) a typical triangular waveform demonstrating the cycling of potential with time, (b) the current response with time, and (c) the current-potential representations (c).

Various parameters can be adjusted including the scan rate, the number of scan cycles, and the direction of the potential scan i.e., whether the forward scan moves from positive to negative voltages or vice versa. For publication, data is typically collected at a scan rate of 20 mV/s with at least 3 scan cycles.

Cyclic voltammetry measurements

Cyclic voltammetry (CV) is one type of potentiodynamic electrochemical measurements. Generally speaking, the operating process is a potential-controlled reversible experiment, which scans the electric potential before turning to reverse direction after reaching the final potential and then scans back to the initial potential, as shown in Figure 7.4a. When voltage is applied to the system changes with time, the current will change with time accordingly as shown in Figure 7.4b. Thus, the curve of current and voltage, illustrated in Figure 7.4c, can be represented from the data, which can be obtained from Figure 7.4a and b.

Although CV was first practiced using a hanging mercury drop electrode, based on the work of Nobel Prize winner Heyrovský (Figure 7.5), it did not gain widespread until solid electrodes like Pt, Au and carbonaceous electrodes were used, particularly to study anodic oxidations. A major advance was made when mechanistic diagnostics and accompanying quantifications became known through the computer simulations. Now, the application of computers and related software packages make the analysis of data much quicker and easier.

Figure 7.5: Czech chemist and inventor Jaroslav Heyrovský (1890 - 1967).

The components of a CV system

As shown in Figure 7.6, the CV systems are as follows:

- The epsilon includes potentiostat and current-voltage converter. The potentiostat is required for con- trolling the applied potential, and a current-to-voltage converter is used for measuring the current, both of which are contained within the epsilon.
- The input system is a function generator. Parameters, such as scan rate and scan range, may be altered through this part. The output part is a computer screen, which can show data and curves directly.
- All electrodes must be immersed in the electrolyte solution.
- Sometimes, the oxygen and water in the atmosphere will dissolve in the solution and will be deoxidized or oxidized when voltage is applied. Therefore, the data will be less accurate. To prevent this from happening, bubbling of an inert gas (nitrogen or argon) is required.
- The key component of the CV systems is the electrochemical cell, which is connected to the epsilon components. The electrochemical cell contains three electrodes: counter electrode (C), working electrode (W) and reference electrode (R). All of them must be immersed in an electrolyte solution when working.

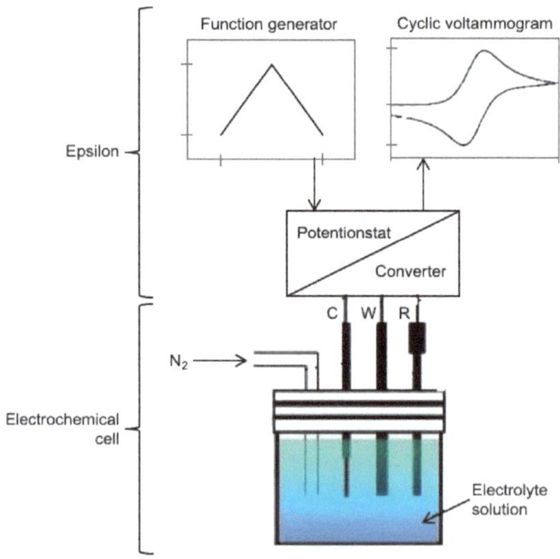

Figure 7.6: Components of cyclic voltammetry systems. Adapted from D. K. Gosser, Jr., *Cyclic Voltammetry Simulation and Analysis of Reaction Mechanisms*, Wiley-VCH, New York (1993).

In order to better understand the electrodes mentioned above, three kinds of electrodes will be discussed in more detail.

- Counter electrodes (C) are non-reactive high surface area electrodes, for which platinum gauze is the common choice (Figure 7.7).
- The working electrode (W) is commonly an inlaid disc electrode (Pt, Au, graphite, etc.) of well-defined area (Figure 7.8). Other geometries may be available in appropriate circumstances, such as hanging mercury drop electrode (Figure 7.9) or dropping mercury electrode (Figure 7.10), cylinder, band, arrays, and grid electrodes.
- For the reference electrode (R), aqueous Ag/AgCl or calomel half cells are commonly used and can be obtained commercially or easily prepared in the laboratory. Sometimes, a simple silver or platinum wire is used in conjunction with an internal potential reference provided by ferrocene (Figure 7.11), when a suitable conventional reference electrode is not available. Ferrocene undergoes a one-electron oxidation at a low potential, around 0.5 V versus a saturated calomel electrode (SCE). It is also been used as standard in electrochemistry as $F^+/F = 0.64$ V versus a normal hydrogen electrode (NHE).

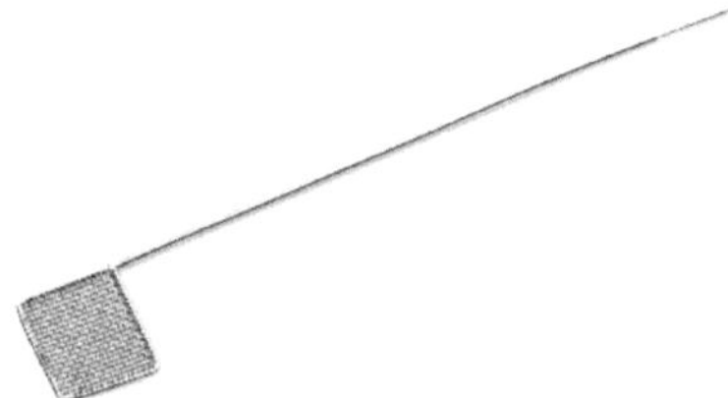

Figure 7.7: A typical platinum gauze electrode.

Figure 7.8: A typical inlaid carbon disc electrodes.

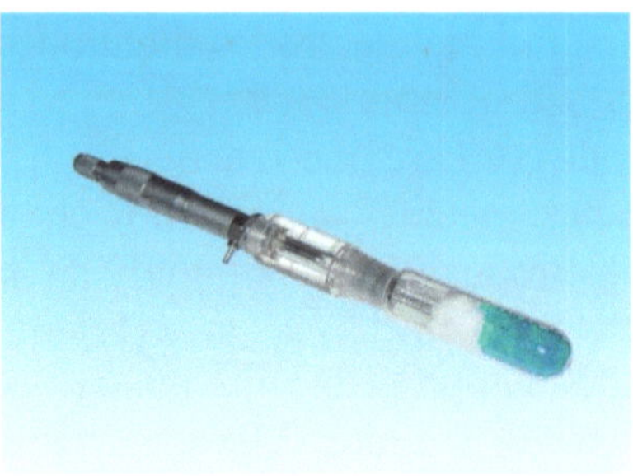

Figure 7.9: A hanging mercury drop electrode.

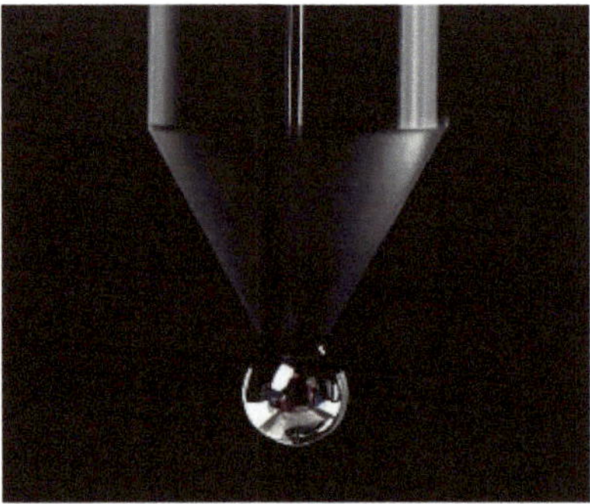

Figure 7.10: A dropping mercury electrode.

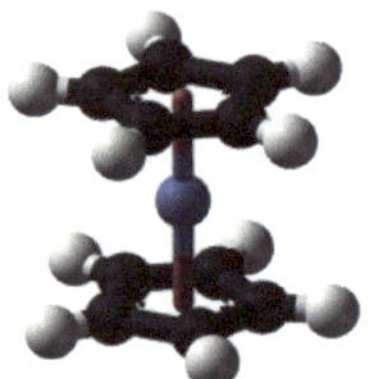

Figure 7.11: The structure of $Fe(\eta^5\text{-}C_5H_5)_2$ (ferrocene).

Cyclic voltammetry systems employ different types of potential waveforms (Figure 7.12) that can be used to satisfy different requirements. Potential waveforms reflect the way potential is applied to this system. These different types are referred to by characteristic names, for example, cyclic voltamme-try, and differential pulse voltammetry. The cyclic voltammetry analytical method is the one whose potential waveform is generally an isosceles triangle (Figure 7.12a).

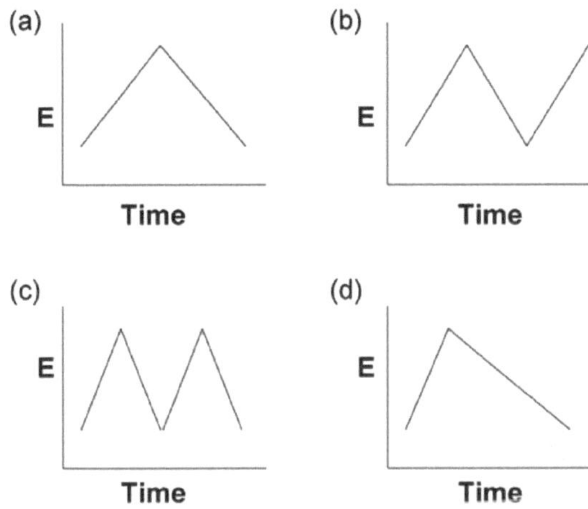

Figure 7.12: Examples of different waveforms of CV systems, illustrating various possible cycles.

The physical principles on which the CV systems are based

As mentioned above, there are two main parts of a CV system: the electrochemical cell and the epsilon. Figure 7.13 shows the schematic drawing of circuit diagram in electrochemical cell.

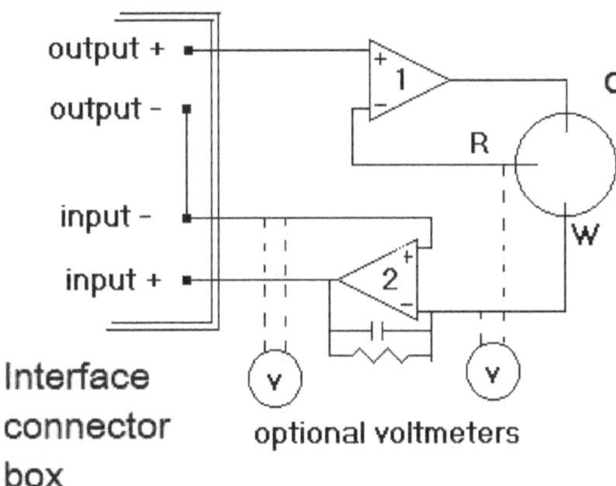

Figure 7.13: Diagram of a typical cyclic voltammetry circuit layout. Adapted from R. G. Compton and C. E. Banks, *Understanding Voltammetry*, World Scientific, Sigapore (2007). Copyright: World Scientific Publishing (2007).

In a voltammetric experiment, potential is applied to a system, using working electrode (W in Figures 7.6 and 7.13) and the reference electrode (R = in Figures 7.6 and 7.13), and the current response is measured using the working electrode and a third electrode, the counter electrode (C in in Figures 7.6 and 7.13). The typical current-voltage curve for ferricyanide/ferrocyanide,

$$[Fe(CN)_6]^{3-} + e^- \rightarrow [Fe(CN)_6]^{4-}$$

is shown in Figure 7.14.

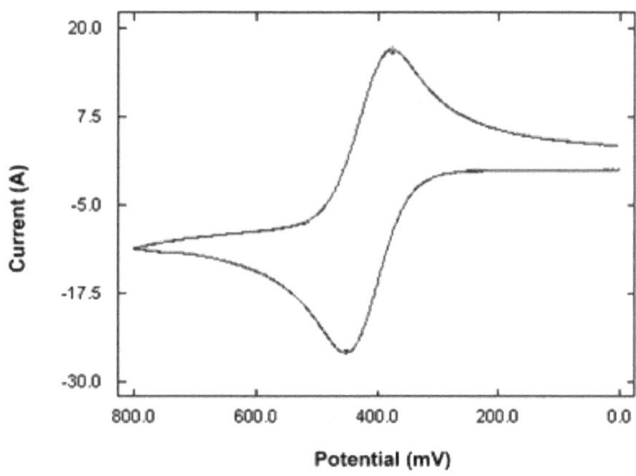

Figure 7.14: Typical curve of I-V curve for ferricyanide/ferrocyanide.

Reading a voltammogram

From a cyclic voltammetry experiment, a graph called a voltammogram will be obtained. Because both the oxidation and reduction half-reactions occur at the working electrode surface, steep changes in the current will be observed when either of these half-reactions occur. A typical voltammogram will feature two peaks, where one peak corresponds to the oxidation half-reaction and the other to the reduction half-reaction. In an oxidation half-reaction in an electrochemical cell, electrons flow from the species in solution to the electrode resulting in an anodic current, i_a. Frequently, this oxidation peak appears when scanning from negative to positive potentials (Figure 7.14). In a reduction half-reaction in an electrochemical cell, electrons flow from the electrode to the species in solution, resulting in a cathodic current, i_c. This type of current is most often observed when scanning from positive to negative potentials (Figure 7.14). When the starting reactant is completely oxidized or completely reduced, peak anodic current, i_{pa}, and peak cathodic current, i_{pc}, respectively, are reached. Then, the current decays as the oxidized or reduced species leaves the electrode surface. The shape of these anodic and cathodic peaks can be modeled with the Nernst equation,

$$E_{eq} = E^{\circ\prime} + (0.059/n) \log [reactant]/[product]$$

where n is the number of electrons transferred and $E^{\circ\prime}$ is the formal reduction potential,

$$E^{\circ\prime} = (E_{pa} + E_{pc})/2$$

Important values from the voltammogram

Several key pieces of information can be obtained through examination of the voltammogram including i_{pa}, i_{pc}, and the anodic and cathodic peak potentials. i_{pa} and i_{pc} both serve as important measures of catalytic activity: the larger the peak currents, the greater the activity of the catalyst. Values for i_{pa} and i_{pc} can be obtained through one of two methods: physical examination of the graph or the Randles-Sevick equation. To determine the peak potentials directly from the graph, a vertical tangent line from the peak current is intersected with an extrapolated baseline. In contrast, the Randles-Sevick equation uses information about the electrode and the experimental parameters to calculate the peak current,

$$i_p = (2.69 \times 10^5)n^{3/2}AD^{1/2}Cv^{1/2}$$

where A = electrode area, D = diffusion coefficient, C = concentration, and v = scan rate.

Anodic peak potential, E_{pa}, and cathodic peak potential, E_{pc}, can also be obtained from the voltammogram by determining the potential at which ipa and ipc respectively occur. These values are an indicator of the relative magnitude of the reaction rate. If the exchange of electrons between the oxidizing and reducing agents is fast, they form an electrochemically reversible couple. These redox couples fulfill the relationship:

$$\Delta E_p = E_{pa} \times E_{pc} \equiv 0.059/n$$

In contrast, a nonreversible couple will have a slow exchange of electrons,

$$\Delta E_p > 0.059/n$$

However, it is important to note that ΔE_p is dependent on scan rate.

Analysis of reaction kinetics

The Tafel and Butler-Volmer equations allow for the calculation of the reaction rate from the current-potential data generated by the voltammogram. In these analyses, the rate of the reaction can be expressed as two values: k° and i_o. The standard rate constant, k°, is a measure of how fast the system reaches equilibrium: the larger the value of k°, the faster the reaction. The exchange current density, (i_o) is the current flow at the surface of the electrode at equilibrium: the larger the value of i_o, the faster the reaction. While both i_o and k° can be used, i_o is more frequently used because it is directly related to the overpotential through the current-overpotential and Butler-Volmer equations. When the reaction is at equilibrium, k° and i_o are related by,

$$i_o = nFk^\circ C_{o,eq}^{(1-a)} C_{R,eq}^{a}$$

where $C_{O,eq}$ and $C_{R,eq}$ are the equilibrium concentrations of the oxidized and reduced species, respectively, and a is the symmetry factor.

Tafel equation

In its simplest form, the Tafel equation is expressed as,

$$E - E^\circ = a + b \log(i)$$

where a and b can be a variety of constants. Any equation which has this form is considered a Tafel equation. For example, the relationship between current, potential, the concentration of reactants and products, and k can be expressed as,

$$C_O(0,t) - C_R(0,t)e^{\{[nF/RT](E-E')\}} = [i/nF\ k^\circ]\ [e^{\{[anF/RT](E-E')\}}]$$

where $C_O(0,t)$ and $C_R(0,t)$ = concentrations of the oxidized and reduced species, respectively, at a specific reaction time, F = Faraday constant, R = gas constant, and T = temperature. At very large overpotentials, this equation reduces to a Tafel equation,

$$E-E^\circ = [RT/(1-a)nF]\ln(i) - [RT/(1-a)nF]\ln(i_o)$$

where

$$a = -[RT/(1-a)nF]\ln(i_O)$$

$$b = [RT/(1-a)nF]$$

The linear relationship between $E-E^\circ$ and $\log(i)$ can be exploited to determine i_O through the formation of a Tafel plot (Figure 7.15), E-E versus $\log(i)$. The resulting anodic and cathodic branches of the graph have slopes of $[(1-a)nF/2.3RT]$ and $[-anF/2.3RT]$, respectively. An extrapolation of these two branches results in a y-intercept = $\log(i_O)$. Thus, this plot directly relates potential and current data collected by cyclic voltammetry to i_O.

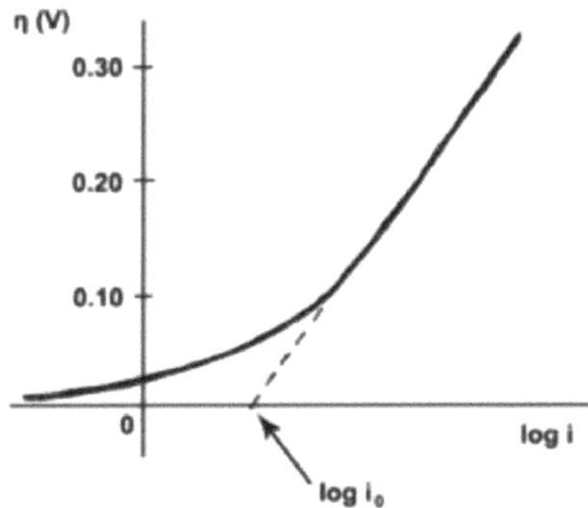

Figure 7.15: Example of an idealized Tafel plot of η (V) verses log i. Reprinted with the permission of R. C. M. Jakobs under the GNU Free Documentation License (2010).

Butler-Volmer equation

While the Butler-Volmer equation resembles the Tafel equation, and in some cases can even be reduced to the Tafel formulation, it uniquely provides a direct relationship between i_o and H. Without simplification, the Butler-Volmer equation is known as the current-overpotential,

$$i/i_o = [C_O (0,t)/C_{o,eq}]\, e^{\{[anF/RT](E-E^\circ)\}} - [C_R (0,t)/C_{R,eq}]\, e^{\{[(1-a)nF/RT](E-E^\circ)\}}$$

If the solution is well-stirred, the bulk and surface concentrations can be assumed to be equal and the equation can be reduced to the Butler-Volmer equation,

$$I = i_o[e^{\{[anF/RT](E-E^\circ)\}} - e^{\{[(1-a)nF/RT](E-E^\circ)\}}]$$

Applications of cyclic voltammetry

Oxygen reduction reaction (ORR) catalysis

The oxygen reduction reaction (ORR) is also the most important reaction in life processes such as biological respiration and is the key cathodic process in fuel cells. ORR in aqueous solutions occurs mainly by two pathways: direct 4-electron reduction pathway from O_2 to H_2O,

$$O_2\,(g) + 4H^+\,(aq) + 4e^- \rightarrow 2H_2O\,(l) \qquad E^0 = +1.229\ V$$

and 2-electron reduction pathway from O_2 to hydrogen peroxide (H_2O_2),

$$O_2\,(g) + 2H^+\,(aq) + 2e^- \rightarrow H_2O_2\,(l) \qquad E^0 = +0.670\ V$$

In non-aqueous aprotic solvents and/or in alkaline solutions, the 1-electron reduction pathway from O_2 to superoxide (O_2^-) can also occur.

While the issue of a slow oxygen reduction reaction (ORR) reaction rate has been addressed in many ways, it is most often overcome with the use of catalysts.

Platinum catalysis

Traditionally, platinum catalysts have demonstrated the best performance at 30 °C, the ORR i_O on a Pt catalyst is 2.8×10^{-7} A/cm^2 compared to the limiting case of ORR where $i_O = 1 \times 10^{-7}$ A/cm^2. Pt is particularly effective as a catalyst for the ORR in PEMFCs because its binding energy for both O and OH is the closest to ideal of all the bulk metals, its activity is the highest of all the bulk metals, its selectivity for O_2 adsorption is close to 100%, and its extreme stability under a variety of acidic and basic conditions as well as high operating voltages Figure 7.16.

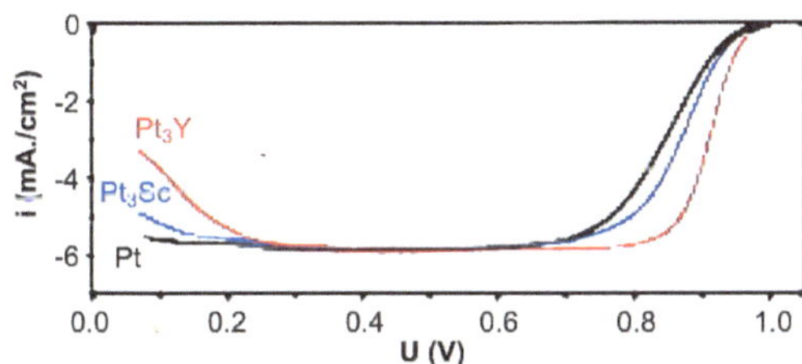

Figure 7.16: Anodic sweeps of cyclic voltammograms of Pt, Pt₃Sc, and Pt₃Y in 0.1 M HClO₄ at 20 mV/s. Adapted from J. Greeley, I. E. L. Stephens, A. S. Bondarenko, T. P. Johansson, H. A. Hansen, T. F. Jaramillo, J. Rossmeisl, I. Chorkendor , and J. K. Nørskov, Alloys of platinum and early transition metals as oxygen reduction electrocatalysts. *Nat. Chem.*, 2009, 1, 552. Copyright Springer Nature (2009).

Metal-nitrogen-carbon composite catalysts

Non-precious metal catalysts (NPMCs) show great potential to reduce the cost of the catalyst without sacrificing catalytic activity. The best NPMCs currently in development have comparable or even better ORR activity and stability than platinum-based catalysts in alkaline electrolytes; in acidic electrolytes, however, NPMCs perform significantly worse than platinum-based catalysts.

In particular, transition metal-nitrogen-carbon composite catalysts (M-N-C) are the most promising type of NPMC. The highest-performing members of this group catalyze the ORR at potentials within 60 mV of the highest-performing platinum catalysts. Additionally, these catalysts have excellent stability: after 700 hours at 0.4 V, they do not show any performance degradation. In a comparison of high-performing PANI-Co-C and PANI-Fe-C (PANI = polyaniline, Figure 7.17), cyclic voltammetry was used to compare the activity and performance of these two catalysts in H_2SO_4. The Co-PANI-C catalyst was found to have no reduction-oxidation features on its voltammogram whereas Fe-PANI-C was found to have two redox peaks at ~0.64. These Fe-PANI-C peaks have a full width at half maximum of ~100 mV, which is indicative of the reversible one-electron Fe^{3+}/Fe^{2+} reduction-oxidation (theoretical FWHM = 96 mV). It was also determined the exchange current density using the Tafel analysis and found that Fe-PANI-C has a significantly greater i_O ($i_O = 4\times10^{-8}$ A/cm^2) compared to Co-PANI-C ($i_O = 5 \times10^{-10}$ A/cm^2). These differences not only demonstrate the higher ORR activity of Fe- PANI-C when compared to Co-PANI-C, but also suggest that the ORR-active sites and reaction mechanisms are different for these two catalysts. While the structure of Fe-PANI-C has been examined the structure of Co-PANI-C is still being investigated.

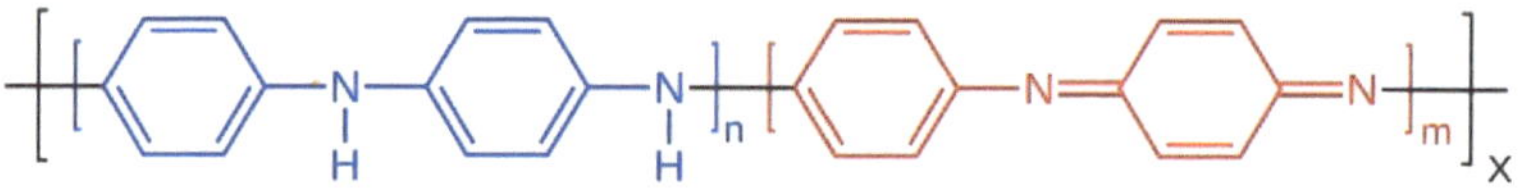

Figure 7.17: The structure of polyaniline.

While the majority of the M-N-C catalysts show some ORR activity, the magnitude of this activity is highly dependent upon a variety of factors; cyclic voltammetry is critical in the examination of the relationships between each factor and catalytic activity.

Proton exchange membranes

Proton exchange membrane fuel cells (PEMFCs) are one promising alternative to traditional combustion engines, that takes advantage of the exothermic hydrogen oxidation reaction in order to generate energy and water (Table 7.3).

	Acidic electrolyte	**Acidic redox potential at STP (V)**	**Basic electrolyte**	**Basic redox potential at STP (V)**
Anode half-reaction	$2H_2 \rightarrow 4H^+ + 4e^-$		$2H_2 + 4OH^- \rightarrow$ $4H_2O + 4e^-$	
Cathode half-reaction	$O_2 + 4e^- + 4H^+ \rightarrow$ $2H_2O$	1.23	$O_2 + 4e^- + 2H_2O \rightarrow$ $4OH^-$	0.401

Table 7.3 Summary of oxidation-reduction reactions in a PEMFC in acidic and basic electrolytes.

The basic PEMFC consists of an anode and a cathode separated by a proton exchange membrane (Figure 7.18). This membrane is a key component of the fuel cell because for the redox couple reactions to successfully occur, protons must be able to pass from the anode to the cathode. The membrane in a PEMFC is usually composed of Nafion™ (Figure 7.19), which is a polyfluorinated sulfonic acid, and exclusively allows protons to pass through. As a result, electrons and protons travel from the anode to the cathode through an external circuit and through the proton exchange membrane, respectively, to complete the circuit and form water.

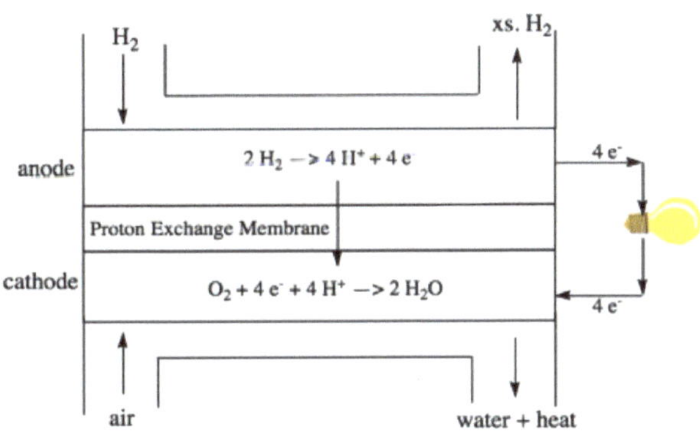

Figure 7.18: Schematic of a proton exchange membrane fuel cell (PEMFCs).

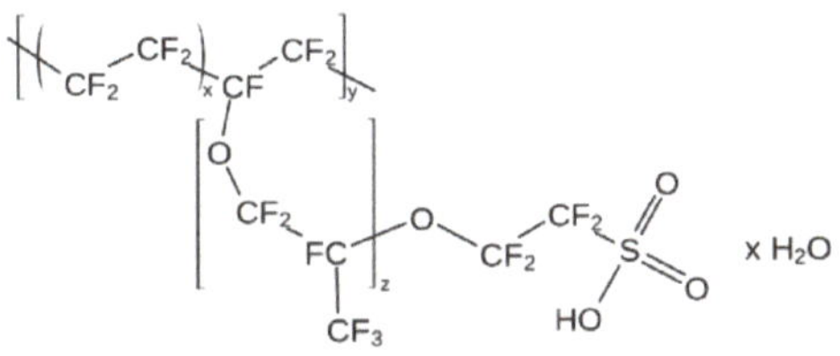

Figure 7.19. The structure of the sulfonated tetrafluoroethylene-based fluoropolymer-copolymer, Nafion™.

PEMFCs present many advantages compared to traditional combustion engines. They are more efficient and have a greater energy density than traditional fossil fuels. Additionally, the fuel cell itself is very simple with few or no moving parts, which makes it long-lasting, reliable, and very quiet. Most importantly, however, the operation of a PEMFC results in zero emissions as the only byproduct is water (Table 7.4). However, the use of PEMFCs has been limited because of the slow reaction rate for the oxygen reduction half-reaction (ORR). Reaction rates, k^o, for reduction-oxidation reactions such as these tend to be on the order of 10^{-10} - 10^{-9}, where 10^{-10} is the fastest reaction rate and 10^{-9} is the slowest reaction rate. Compared to the hydrogen oxidation half-reaction (HOR), which has a reaction rate of $k^o = 1 \times 10^{-10}$ cm/s, the reaction rate for the ORR is $k^o \sim 1 \times 10^{-9}$ cm/s. Thus, the ORR is the kinetic rate-limiting half-reaction and its reaction rate must be increased for PEMFCs to be a viable alternative to combustion engines. Because cyclic voltammetry can be used to examine the kinetics of the ORR reaction, it is a critical technique in evaluating potential solutions to this problem.

Advantages	Disadvantages
More efficient than combustion engines	ORR half-reaction too slow for commercial use
Greater energy density than fossil fuels	Hydrogen fuel is not readily available
Long-lasting	Water circulation must be managed to keep the pro- ton exchange membrane hydrated
Reliable	
Quiet	
No harmful emissions	

Table 7.4: Summary of advantages and disadvantages of PEMFCs as an alternative to combustion engines.

Solar cell materials

The band gap of a semiconductor is a very important value to be determined for photovoltaic materials. Figure 7.20 shows the relative energy level involved in light harvesting of an organic solar cell. The energy difference (E_g) between the lowest unoccupied molecular orbital (LUMO) and the highest occupied molecular orbital (HOMO), which determines the efficiency. The oxidation and reduction of an organic molecule involve electron transfers (Figure 7.21), and CV measurements can be used to determine the potential change during redox. Through the analysis of data obtained by the CV measurement the electronic band gap is obtained.

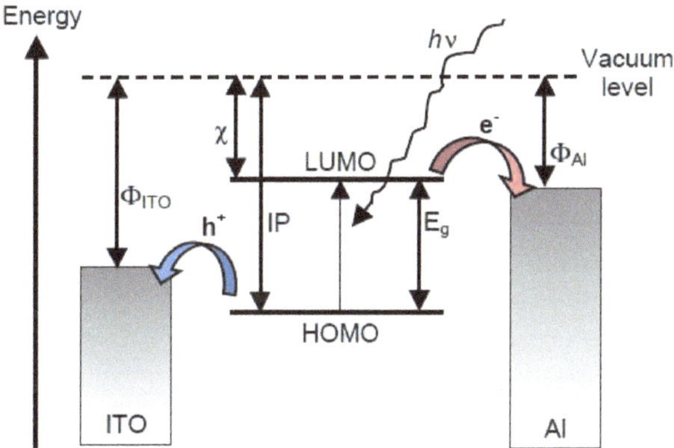

Figure 7.20: Diagram showing energy level and light harvesting of and organic solar cell. Adapted from S. B. Darling, Block copolymers for photovoltaic. *Energy Environ. Sci.*, 2009, 2, 1266. Copyright: Royal Society of Chemistry (2009).

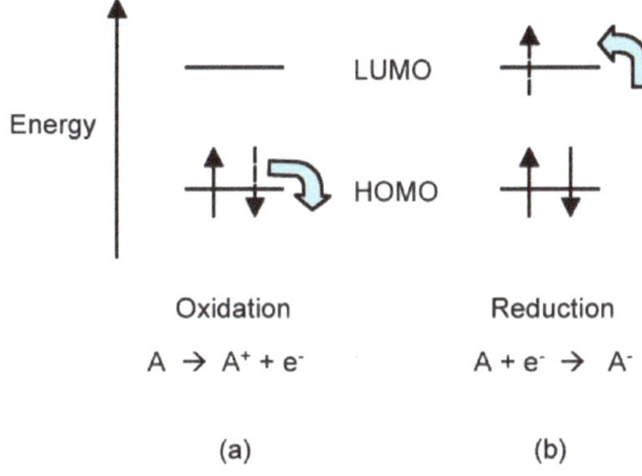

Figure 7.21: Diagram showing energy level and light harvesting of organic solar cell. Adapted from D. K. Gosser, Jr., *Cyclic Voltammetry Simulation and Analysis of Reaction Mechanisms*, Wiley-VCH, New York (1993). Copyright: Wiley-VCH (1993).

Chronocoulometry

Theory

Chronocoulometry, as indicated by the name, is a technique in which the charge is measured (i.e., *coulometry*) as a function of time (i.e., *chrono*). There are various types of coulometry. The one discussed here is potentiostatic coulometry in which the potential (or voltage) is set and, as a result, charge flows through the cell. The input and output example graphs can be seen in Figure 7.22. The input is a potential step that spans the reduction potential of the electroactive species. If this potential step is performed in an electro- chemical cell that does not contain and electroactive species, only capacitive current will flow (Figure 7.23), in which the ions migrate in such a way that charges are aligned (positive next to negative), but no charge is transferred. Once an electroactive species is introduced into the system however, the faradaic current begins to flow. This current is a result of the electron transfer between the electrode and the electroactive species.

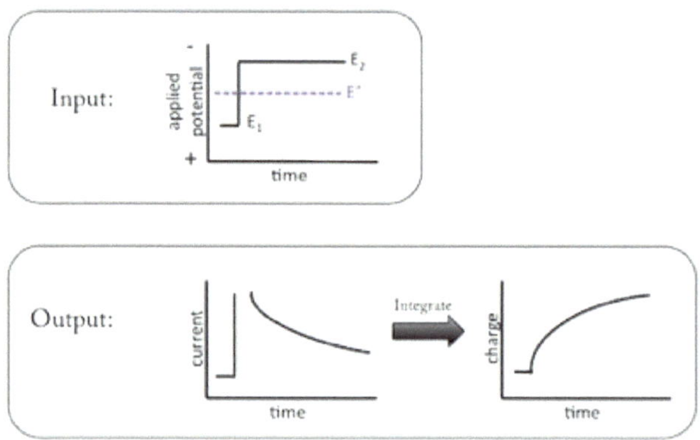

Figure 7.22: Input potential step (a) and output charge transfer (b) as used in chronocoulometry.

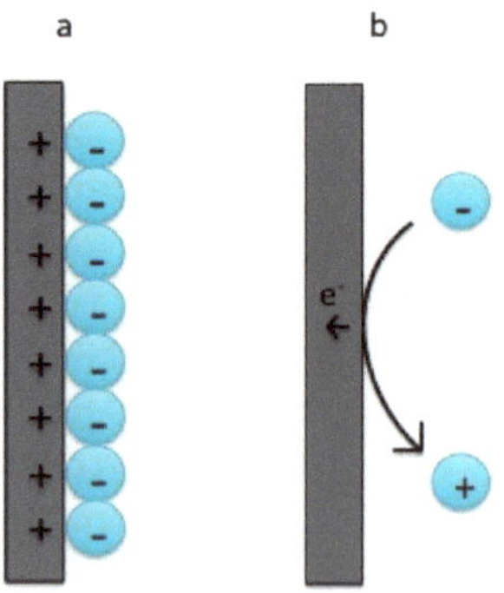

Figure 7.23: Capacitive alignment (a) and faradaic charge transfer (b) the two sources of current in an electrochemical cell.

Electroplating

Electroplating is an electrochemical process that utilizes techniques such as chronocoulometry to electrode- posit a charged chemical from a solution as a neutral chemical on the surface of another chemical. These chemicals are typically metals. The science of electroplating dates back to the early 1800s when Luigi Valentino Brugnatelli (Figure 7.24) electroplated gold from solution onto silver metals. By the mid 1800s, the process of electroplating was patented by cousins George and Henry Elkington (Figure 7.25). The Elkingtons brought electroplated goods to the masses by producing consumer products such as artificial jewelry and other commemorative items (Figure 7.26).

Figure 7.24: Portrait of Italian chemist and inventor Luigi Valentino Brugnatelli (1761 - 1818) who discovered the process for electroplating in 1805.

Figure 7.25: Portrait of English businessman George Elkington (1801 - 1865) who patented the first commercial electroplating process.

Figure 7.26: A commemorative inkstand gilded using the process of electro-plating.

Bibliography

R. N. Adams, *Electrochemistry at Solid Electrodes*, Marcel Dekker, New York (1968).

A. J. Bard and L. R. Faulkner, *Electrochemical Methods: Fundamentals and Applications*, Wiley (2000).

R. G. Compton, and C. E. Banks, *Understanding Voltammetry*, World Scientific, Sigapore, (2007).

N. Daems, X. Sheng, Y. Alvarez-Gallego, I. F. J. Vankelecom, and P. P. Pescarmona, Iron-containing N-doped carbon electrocatalysts for the cogeneration of hydroxylamine and electricity in a H_2-NO fuel cell. *Green Chem.*, 2016, **18**, 1547.

S. B. Darling, Block copolymers for photovoltaic. *Energy Environ. Sci.*, 2009, **2**, 1266.

Q. Dong, S. Santhanagopalan, and R.E. White, Simulation of the oxygen reduction reaction at an RDE in 0.5 M H_2SO_4 including an adsorption mechanism. *J. Electrochem. Soc.*, 2007, **154**, A888.

S. W. Feldberg, *A General Method for Simulation*, Vol. 3 in Electroanalytical Chemistry Series, Marcel Dekker, NewYork (1969).

J. Greeley, I. E. L. Stephens, A. S. Bondarenko, T. P. Johansson, H. A. Hansen, T. F. Jaramillo, J. Rossmeisl, I. Chorkendor, and J. K. Nørskov, Alloys of platinum and early transition metals as oxygen reduction electrocatalysts. *Nat. Chem.*, 2009, **1**, 552.

O. T. Holton and J. W. Stevenson, The role of platinum in proton exchange membrane fuel cells. *Platinum Metals Rev.*, 2013, **57**, 259.

W. Kemula and Z. Kublik, Observation of transient intermediates in redox processes by variable voltage oscillo-polarography and cyclic voltammetry. *Nature*, 1958, **182**, 793.

P. T. Kissinger and W. R. Heineman, Cyclic voltammetry. *J. Chem. Educ.*, 1981, **60**, 702.

J. Larminie and A. Dicks, *Fuel Cell Systems Explained*, 2nd edn., John Wiley & Sons, Inc., Hoboken (2003).

A. Mendez, L. E. Moron, L Ortiz-Frade, Y. Meas, R Ortega-Borges, G. Trejo, Thermodynamic studies of PEG (M_w 20,000) adsorption onto a polycrystalline gold electrode. *J. Electrochem. Soc.*, 2011, **158**, F45.

R. S. Nicholson and I. Shain, Theory of stationary electrode polarography. single scan and cyclic methods applied to reversible, irreversible, and kinetic systems. *Anal. Chem.*, 1964, **36**, 706.

L. Piszczek, A. Lgnatowicz, J. Kielbase, Cyclic voltammetry. *J. Chem. Edu.*, 1998, **65**, 171.

C. Song and J. Zhang, *Electrocatalytic Oxygen Reduction Reaction in PEM Fuel Cell Electrocatalysts and Catalyst Layers*, Springer, London (2008).

G. Wu, K. More, C. Johnston, and P. Zelenay, High-performance electrocatalysts for oxygen reduction derived from polyaniline, iron, and cobalt. *Science*, 2011, **332**, 443.

G. Wu, C. M. Johnston, N. H. Mack, K. Artyushkova, M. Ferrandon, M. Nelson, J. S. Lezama-Pacheco, S. D. Conradson, K. L More, D. J. Myers, and P. Zelenay, Synthesis–structure–performance correlation for polyaniline–Me–C non-precious metal cathode catalysts for oxygen reduction in fuel cells. *J. Mater. Chem.*, 2011, **21**, 11392.

J. Zhu, Z. Zhao, D. Xiao, J. Li, X. Yang, and Y. Wu, Application of cyclic voltammetry in heterogeneous catalysis: NO decomposition and reduction. *Electrochem. Commun.*, 2005, **7**, 58.

116

Chapter 8: Thermogravimetric Analysis

Wala Algozeeb, Caoimhe de Fréin, Vanessa Espinoza,
Nikolaos Soultanidis, Pavan M. V. Raja and Andrew R. Barron

Introduction

Broadly speaking thermal analysis is the study of materials as they change when varying temperatures are applied to them. Such techniques include thermogravimetric analysis (TGA), differential thermal analysis (DTA), and differential scanning calorimetry (DSC).

The main function of TGA is the monitoring of the thermal stability of a material by recording the change in weight of the sample with respect to temperature. Upon heating the sample, any mass loss is recorded to get a final plot of weight versus temperature. Then the results are analyzed to assign the weight loss/gain to a certain chemical/physical change.

Thermogravimetric analysis (TGA) is a technique widely used for determining the decomposition, boiling, and sublimation points for various organic and inorganic materials. In addition, it can be used to calculate the composition of materials based upon physical (e.g., dehydration, desorption) and chemical (decomposition) changes. The sample to be analyzed is placed into platinum or alumina pan, and along with an empty pan as a standard, is placed onto two high precision balances inside a high temperature oven. Its basic rule of function is the high precision measurement of weight gain/loss with increasing temperature under inert or reactive atmospheres. Each weight change corresponds to physical (sublimation, adsorption/desorption) or chemical (oxidation, pyrolysis, etc.) processes that take place by increasing the temperature.

History

The exploration of thermal analysis techniques first began in 1887 with Le Chatelier (Figure 8.1) where he focused on the behavior of clays upon heating. This initial thermal experimentation allowed for further exploration into thermal analysis, which eventually led to the development of thermogravimetric analysis, in 1915 by Japanese physicist Honda (Figure 8.2).

Figure 1.1: French chemist Henry Louis Le Chatelier (1850 - 1936).

Figure 8.2: Japanese physicist Kotaro Honda (1870 - 1954).

Instrumentation and experimental set up

A thermal gravimetric analyzer (TGA) records the change in weight for a sample of known weight as temperature increases under a controlled atmosphere. The main components of a TGA (Figure 8.3) are:

- micro-sensitive balance,
- furnace,
- sample chamber,
- a purge system.

Inside the TGA (Figure 8.4), there are two pans, a reference pan and a sample pan. The pan material can be either aluminum or platinum. The type of pan used depends on the maximum temperature of a given run. As platinum melts at 1760 °C and aluminum melts at 660 °C, platinum pans are obviously chosen when the maximum temperature exceeds 660 °C. Under each pan there is a thermocouple which reads the temperature of the pan.

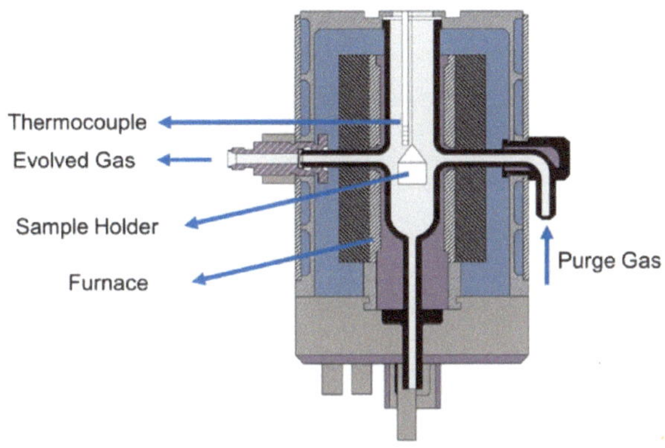

Figure 8.3: The main components of a typical TGA instrument.

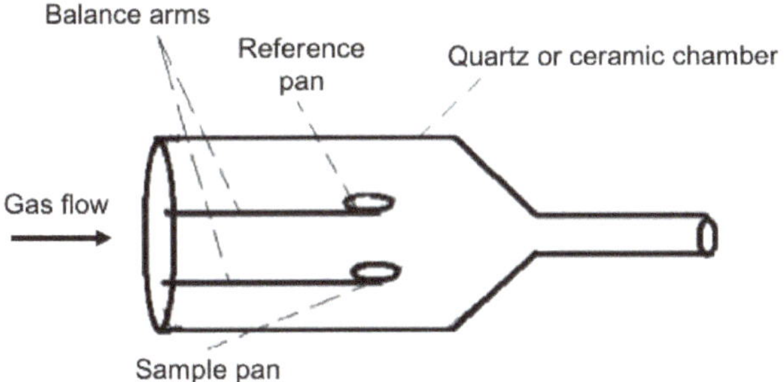

Figure 8.4: Schematic representation of a TGA apparatus.

Before the start of each run, each pan is balanced on a balance arm. The balance arms should be calibrated to compensate for the differential thermal expansion between the arms. If the arms are not calibrated, the instrument will only record the temperature at which an event occurred and not the change in mass at a certain time. To calibrate the system, the empty pans are placed on the balance arms and the pans are weighed and zeroed.

The TGA is then heated in a controlled manner and the absolute weight of a single pan and/or differential weight of the sample and reference pans is measured. Typically, the sample mass range should be between 0.1 - 10 mg and the heating rate should be 3 - 5 °C/min.

As well as recording the change in mass, the heat flow into the sample pan (differential scanning calorimetry, DSC) can also be measured and the difference in temperature between the sample and reference pan (differential thermal analysis, DTA). DSC is quantitative and is a measure of the total energy of the system. This is used to monitor the energy released and absorbed during a chemical reaction for a changing temperature. The DTA shows if and how the sample phase changed. If the DTA is constant, this means that there was no phase change. Figure 8.5 shows a DTA with typical examples of an exotherm and an endotherm.

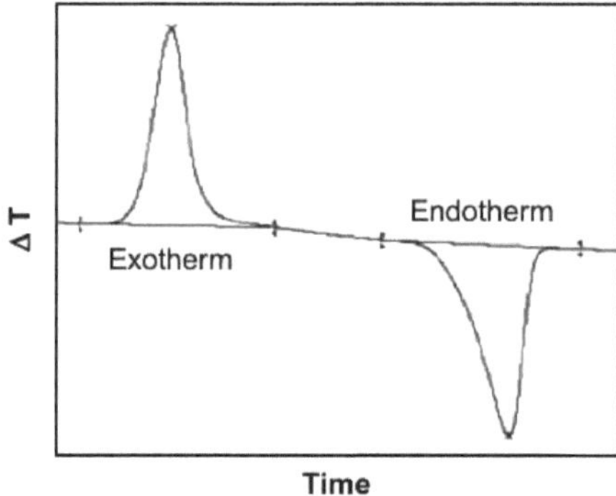

Figure 8.5: Simplified representation of the DTA (ΔT versus time) for an exotherm and an endotherm.

When the sample melts, the DTA dips which signifies an endotherm. When the sample is melting it requires energy from the system. Therefore, the temperature of the sample pan decreases compared with the temperature of the reference pan. When the sample has melted, the temperature of the sample pan increases as the sample is releasing energy. Finally, the temperatures of the reference and sample pans equilibrate resulting in a constant DTA. When the sample evaporates, there is a peak in the DTA. This exotherm can be explained in the same way as the endotherm.

Advantages of TGA

- Very sensitive to weight fluctuation ($\pm 0.01\%$).
- Ability to analyze weight change under different conditions, e.g., argon, air, etc.
- Can be programmed to perform multiple heating cycles.

- TGA can be combined with IR or MS analysis of the volatile products.

Disadvantages of TGA

- Most TGA instruments can only reach temperatures of 1200 °C.
- Imaging techniques are not possible during heating process.
- Sample capacity is relatively low (less that 1000 mg).

Differentiation of physical and chemical processes

Table 8.1 summarizes common physical and chemical changes that occur, and the expected experimental observation on TGA. As may be seen, the combined use of differential thermal analysis (DTA) or differential scanning calorimetry (DSC) is required to fully assign a thermal process.

Process	TGA	DTA/DSC
Melting	No change observed	Endotherm
Boiling/evaporation	Mass loss	Endotherm
Sublimation	Mass loss	Endotherm
Decomposition	Mass loss	Exotherm
Crystallization	No change observed	Exotherm
Adsorption	Mass gain	Exotherm
Absorption	Mass gain	Endotherm
Desorption	Mass loss	Endotherm
Oxidation	Mass gain	Exotherm

Table 8.1: Summary of expected experimental observations for thermogravimetric analysis (TGA) and differential thermal analysis (DTA) or differential scanning calorimetry (DSC) for a range of physical and chemical processes.

Sublimation enthalpy and vapor pressure determination

Metal compounds and complexes are invaluable precursors for the chemical vapor deposition (CVD) of metal and non-metal thin films. In general, the precursor compounds are chosen on the basis of their relative volatility and their ability to decompose to the desired material under a suitable temperature regime. Unfortunately, many readily obtainable (commercially available) compounds are not of sufficient volatility to make them suitable for CVD applications. Thus, a prediction of the volatility of a metal-organic compounds as a function of its ligand identity and molecular structure is desirable in order to determine the suitability of such compounds as CVD precursors. It has been observed that for organic compounds it was determined that a rough proportionality exists between a compound's melting point and sublimation enthalpy; however, significant deviation is observed for inorganic compounds.

Enthalpies of sublimation for metal-organic compounds have been previously determined through a variety of methods, most commonly from vapor pressure measurements using complex experimental systems such as Knudsen effusion, temperature drop microcalorimetry and, more recently, differential scanning calorimetry (DSC). However, the measured values are highly dependent on the experimental procedure utilized. For example, the reported sublimation enthalpy of Al(acac)$_3$ (Figure 8.6, where M = Al) varies from 47.3 - 126 kJ/mol. Thermogravimetric analysis offers a simple and reproducible method for the determination of the vapor pressure of a potential CVD precursor as well as its enthalpy of sublimation.

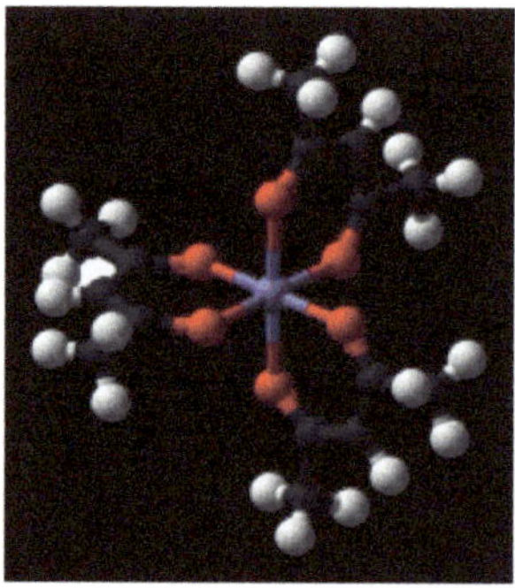

Figure 8.6: Molecular structure of M(acac)$_3$, a typical metal β-diketonate complex.

Determination of sublimation enthalpy

The enthalpy of sublimation is a quantitative measure of the volatility of a particular solid. This information is useful when considering the feasibility of a particular precursor for CVD applications. An ideal sublimation process involves no compound decomposition and only results in a solid-gas phase change, i.e.,

$$[M(L)_n]_{(solid)} \rightarrow [M(L)_n]_{(vapor)}$$

Since phase changes are thermodynamic processes following zero-order kinetics, the evaporation rate or rate of mass loss by sublimation (m_{sub}), at a constant temperature (T), is constant at a given temperature,

$$m_{sub} = \frac{\Delta[mass]}{\Delta t}$$

Therefore, the m_{sub} values may be directly determined from the linear mass loss of the TGA data in isothermal regions. The thermogravimetric and differential thermal analysis of the compound under study is performed to determine the temperature of sublimation and thermal events such as melting. Figure 8.7 shows a typical TG/DTA plot for a gallium chalcogenide cubane compound (Figure 8.8).

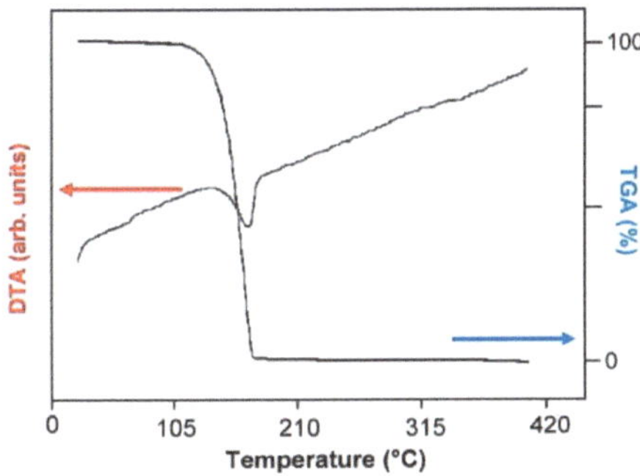

Figure 8.7: A typical thermogravimetric/differential thermal analysis (TG/DTA) analysis of [(EtMe$_2$C)GaSe]$_4$, whose structure is shown in Figure 8.8. Adapted from E. G. Gillan, S. G. Bott, and A. R. Barron, Volatility studies on gallium chalcogenide cubanes: thermal analysis and determination of sublimation enthalpies. *Chem. Mater.*, 1997, 9, 796. Copyright: American Chemical Society (1997).

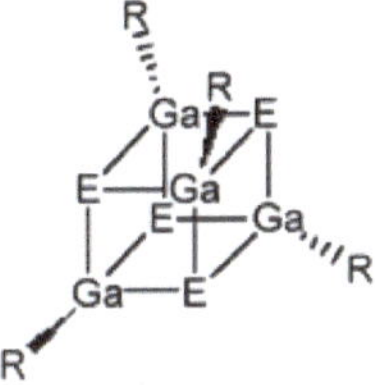

Figure 8.8: Structure of gallium chalcogenide cubane compound, where E = S, Se, and R = CMe$_3$ ('Bu) CMe$_2$Et, CEt$_2$Me, CEt$_3$.

Data collection

In a typical experiment 5 - 10 mg of sample is used with a heating rate of ca. 5 °C/min up to under either a 200-300 mL/min inert (N_2 or Ar) gas flow or a dynamic vacuum (ca. 0.2 Torr if using a typical vacuum pump). The argon flow rate was set to 90.0 mL/min and was carefully monitored to ensure a steady flow rate during runs and an identical flow rate from one set of data to the next.

Once the temperature range is defined, the TGA is run with a preprogrammed temperature profile (Figure 8.9). It has been found that sufficient data can be obtained if each isothermal mass loss is monitored over a period (between 7 and 10 minutes is found to be sufficient) before moving to the next temperature plateau. In all cases it is important to confirm that the mass loss at a given temperature is linear. If it is not, this can be due to either:

- temperature stabilization had not occurred, and so longer times should be spent at each isotherm,
- decomposition is occurring along with sublimation, and lower temperature ranges must be used.

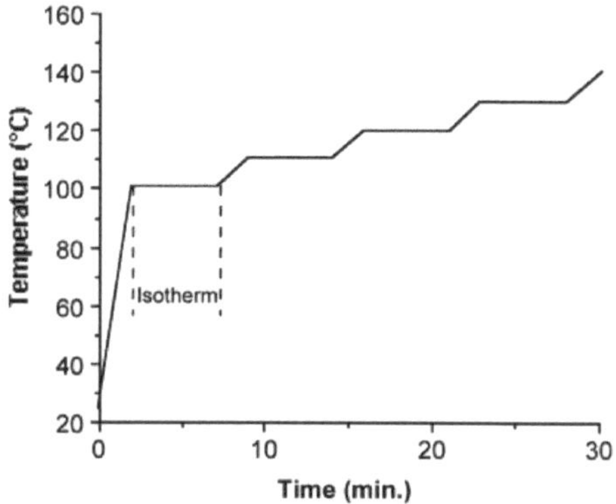

Figure 8.9: A typical temperature profile for determination of isothermal mass loss rate. Adapted from B. D. Fahlman, PhD Thesis, *Chemical Vapor Deposition of Alumina-Based Thin Films*, Department of Chemistry, Rice University (2000).

The slope of each mass drop is measured and used to calculate sublimation enthalpies as discussed below. As an illustrative example, Figure 8.10 displays the data for the mass loss of $Cr(acac)_3$ (Figure 8.6, where M = Cr) at three isothermal regions under a constant argon flow. Each isothermal data set should exhibit a linear relation. As expected for an endothermal phase change, the linear slope, equal to m_{sub}, increases with increasing temperature.

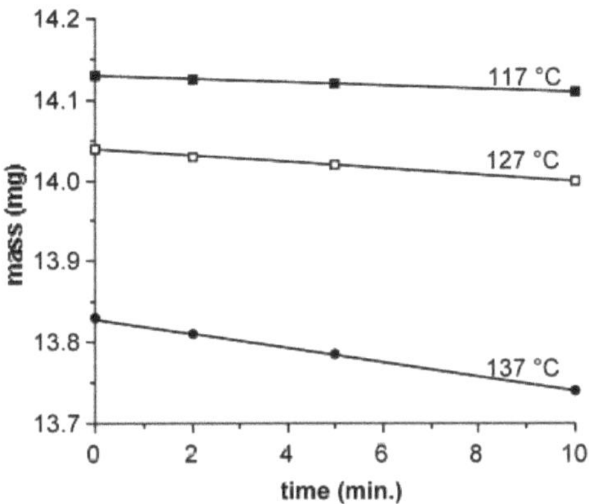

Figure 8.10: Plot of TGA results for Cr(acac)$_3$ performed at different isothermal regions. Adapted from B. D. Fahlman, PhD Thesis, *Chemical Vapor Deposition of Alumina-Based Thin Films*, Department of Chemistry, Rice University (2000).

Samples of iron acetylacetonate (Figure 8.6, where M = Fe) may be used as a calibration standard through ΔH_{sub} determinations before each day of use. If the measured value of the sublimation enthalpy for Fe(acac)$_3$ is found to differ from the literature value by more than 5%, the sample is re-analyzed, and the flow rates are optimized until an appropriate value is obtained. Only after such a calibration is optimized should other complexes be analyzed. It is important to note that while small amounts (<10%) of involatile impurities will not interfere with the ΔH_{sub} analysis, competitively volatile impurities will produce higher apparent sublimation rates.

It is important to discuss at this point the various factors that must be controlled in order to obtain meaningful (useful) m_{sub} data from TGA data. The sublimation rate is independent of the amount of material used but may exhibit some dependence on the flow rate of an inert carrier gas, since this will affect the equilibrium concentration of the compound in the vapor phase. While little variation was observed we decided that for consistency m_{sub} values should be derived from vacuum experiments only.

The surface area of the solid in a given experiment should remain approximately constant; otherwise the sublimation rate (i.e., mass/time) at different temperatures cannot be compared, since as the relative surface area of a given crystallite decreases during the experiment the apparent sublimation rate will also decrease. To minimize this problem, data was taken over a small temperature ranges (ca. 30 °C), and overall sublimation was kept low (ca. 25% mass loss representing a surface area change of less than 15%). In experiments where significant surface area changes occurred the values of m_{sub} deviated significantly from linearity on a $\log(m_{sub})$ versus 1/T plot.

The compound being analyzed must not decompose to any significant degree, because the mass changes due to decomposition will cause a reduction in the apparent m_{sub} value, producing erroneous results. With a simultaneous TG/DTA system it is possible to observe exothermic events if decomposition occurs, however the clearest indication is shown by the mass loss versus time curves which are no longer linear but exhibit exponential decays characteristic of first or second order decomposition processes.

Data analysis

The basis of analyzing isothermal TGA data involves using the Clausius-Clapeyron relation between vapor pressure (p) and temperature (T),

$$\frac{d \ln(p)}{dT} = \frac{\Delta H_{sub}}{RT^2}$$

where ΔH_{sub} is the enthalpy of sublimation and R is the gas constant (8.314 J/K.mol). Since m_{sub} data are obtained from TGA data, it is necessary to utilize the Langmuir equation,

$$p = \left[\frac{2\pi RT}{M_w}\right]^{0.5} m_{sub}$$

that relates the vapor pressure of a solid with its sublimation rate. After integrating the Clausius-Clapeyron relation in log form, substituting in the Langmuir equation, and consolidating the constants, one obtains the useful equality,

$$\log(m_{sub}\sqrt{T}) = \frac{-0.0522(\Delta H_{sub})}{T} + \left[\frac{0.0522(\Delta H_{sub})}{T_{sub}} - \frac{1}{2}\log\left(\frac{1306}{M_w}\right)\right]$$

Hence, the linear slope of a $\log(m_{sub}T^{1/2})$ versus $1/T$ plot yields ΔH_{sub}. An example of a typical plot and the corresponding ΔH_{sub} value is shown in Figure 8.11. In addition, the y intercept of such a plot provides a value for T_{sub}, the calculated sublimation temperature at atmospheric pressure.

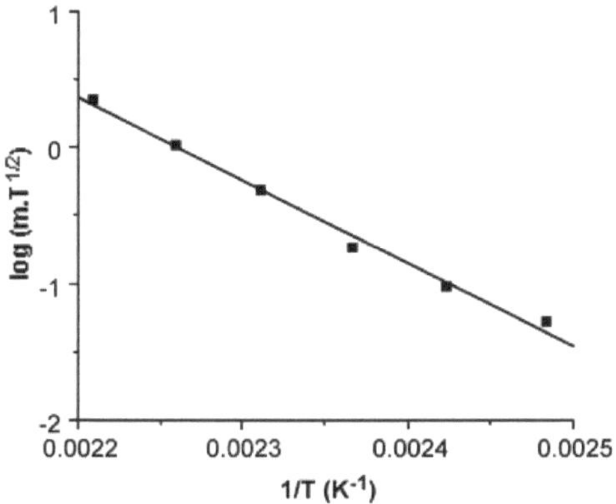

Figure 8.11: Plot of $\log(m_{sub}T^{1/2})$ versus $1/T$ and the determination of the ΔH_{sub} (112.6 kJ/mol) for Fe(acac)₃ ($R^2 = 0.9989$). B. D. Fahlman, PhD Thesis, *Chemical Vapor Deposition of Alumina-Based Thin Films*, Department of Chemistry, Rice University (2000).

Table 8.2 lists the typical results using the TGA method for a variety of metal β-diketonates (Figure 8.12, n = 3), while Table 8.3 lists similar values obtained for gallium chalcogenide cubane compounds.

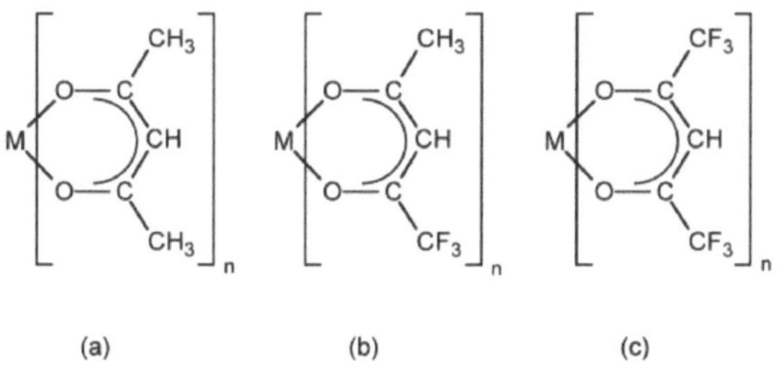

Figure 8.12: Structure of metal β-diketonate complexes: (a) acetylacetonate (acac); (b) trifluoro acetylacetonate (tfac), and (c) hexafluoro acetylacetonate (hfac).

A common method used to enhance precursor volatility and corresponding efficacy for CVD applications is to incorporate partially (Figure 8.12b) or fully (Figure 8.12c) fluorinated ligands. This substitution does results in significant decrease in the ΔH_{sub}, and thus increased volatility (Table 8.2). The observed enhancement in volatility may be rationalized either by an increased amount of intermolecular repulsion due to the additional lone pairs or that the reduced polarizability of fluorine (versus hydrogen) causes fluorinated ligands to have less intermolecular attractive interactions.

Compound	ΔH_{sub} (kJ/mol)	T_{sub} calculated (°C)
Al(acac)$_3$	93	150
Al(tfac)$_3$	74	111
Al(hfac)$_3$	52	70
Cr(acac)$_3$	91	148
Cr(tfac)$_3$	71	109
Cr(hfac)$_3$	46	69
Fe(acac)$_3$	112	161
Fe(tfac)$_3$	96	121
Fe(hfac)$_3$	60	81

Table 8.2: Selected thermodynamic data for metal β-diketonate compounds (Figure 8.12) determined from thermogravimetric analysis. Data from B. D. Fahlman and A. R. Barron, Substituent effects on the volatility of metal β-diketonates. *Adv. Mater. Optics Electron.*, 2000, 10, 223.

Compound	ΔH_{sub} (kJ/mol)	T_{sub} calculated (°C)
[(^{t}Bu)GaS]$_4$	110	94
[(EtMe$_2$C)GaS]$_4$	124	102
[(Et$_2$MeC)GaS]$_4$	137	131
[(Et$_3$C)GaS]$_4$	149	175
[(^{t}Bu)GaSe)]$_4$	119	116
[(EtMe$_2$C)GaSe]$_4$	137	124
[(Et$_2$MeC)GaSe]$_4$	147	136
[(Et$_3$C)GaSe]$_4$	156	189

Table 8.3: Selected thermodynamic data for gallium chalcogenide cubane compounds determined from thermogravimetric analysis. Data from E. G. Gillan, S. G. Bott, and A. R. Barron, Volatility studies on gallium chalcogenide cubanes: thermal analysis and determination of sublimation enthalpies. *Chem. Mater.*, 1997, 9, 796.

Table 8.3 show typical values for gallium chalcogenide cubane compounds. The range observed for gallium chalcogenide cubane compounds ($\Delta S_{sub} = 330 \pm 20$ J/K.mol) is slightly larger than values reported for the metal β-diketonates compounds ($\Delta S_{sub} = 130 - 330$ J/K.mol) and organic compounds (100 - 200 J/K.mol), as would be expected for a transformation giving translational and internal degrees of freedom. For any particular chalcogenide, i.e., $[(R)GaS]_4$, the lowest ΔS_{sub} are observed for the ^{t}Bu (CMe_3) derivatives, and the largest ΔS_{sub} for the CEt_2Me derivatives, see Table 8.3. This is in line with the relative increase in the modes of freedom for the alkyl groups in the absence of crystal packing forces.

Determination of sublimation entropy

The entropy of sublimation, ΔS_{sub}, is readily calculated (Tables 8.4) from the ΔH_{sub} and the calculated T_{sub} data,

$$\Delta S_{sub} = \frac{\Delta H_{sub}}{T_{sub}}$$

Compound	ΔS_{sub} (J/K.mol)
$Al(acac)_3$	220
$Al(tfac)_3$	192
$Al(hfac)_3$	152
$Cr(acac)_3$	216
$Cr(tfac)_3$	186
$Cr(hfac)_3$	134
$Fe(acac)_3$	259
$Fe(tfac)_3$	243
$Fe(hfac)_3$	169

Table 8.4: Calculated entropy of sublimation for metal β-diketonate compounds (Figure 8.12) determined from thermogravimetric analysis. Data from B. D. Fahlman and A. R. Barron, Substituent effects on the volatility of metal β-diketonates. *Adv. Mater. Optics Electron.*, 2000, 10, 223.

Determination of vapor pressure

While the sublimation temperature is an important parameter to determine the suitability of a potential precursor compounds for CVD, it is often preferable to express a compound's volatility in terms of its vapor pressure. However, while it is relatively straightforward to determine the vapor pressure of a liquid or gas, measurements of solids are difficult (e.g., use of the isoteniscopic method) and few laboratories are equipped to perform such experiments. Given that TGA apparatus are increasingly accessible, it provides a simple method for vapor pressure determination.

Substitution of the Langmuir equation allows for the calculation of the vapor pressure (p) as a function of temperature (T). For example, Figure 8.13 shows the calculated temperature dependence of the vapor pressure for $[(^tBu)GaS]_4$. The calculated vapor pressures at 150 °C for metal β-diketonates compounds are given in Table 8.5.

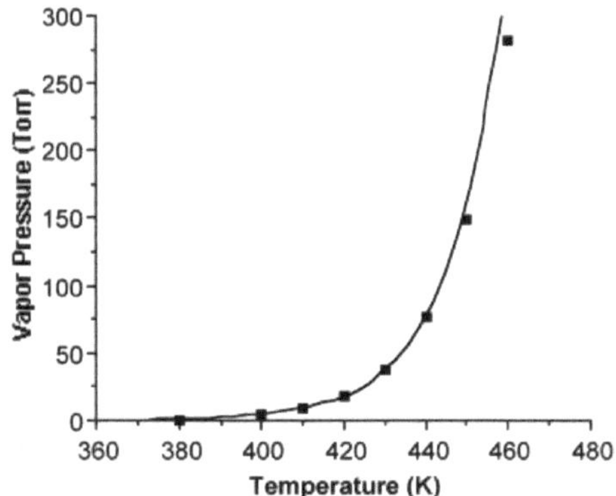

Figure 8.13: A plot of calculated vapor pressure (Torr) against temperature (K) for $[(^tBu)GaS]_4$. Adapted from E. G. Gillan, S. G. Bott, and A. R. Barron, Volatility studies on gallium chalcogenide cubanes: thermal analysis and determination of sublimation enthalpies. *Chem. Mater.*, 1997, 9, 796. Copyright: American Chemical Society (1997).

The TGA approach to show reasonable agreement with previous measurements. For example, while the value calculated for $Fe(acac)_3$ (2.78 Torr @ 113 °C) is slightly higher than that measured directly by the isoteniscopic method (0.53 Torr @ 113 °C); however, it should be noted that measurements using the sublimation bulb method obtained values much lower (8×10^{-3} Torr @ 113 °C). The TGA method offers a suitable alternative to conventional (direct) measurements of vapor pressure.

Compound	Calculated vapor pressure @ 150 °C (Torr)
Al(acac)$_3$	3.261
Al(tfac)$_3$	9.715
Al(hfac)$_3$	29.120
Cr(acac)$_3$	3.328
Cr(tfac)$_3$	9.910
Cr(hfac)$_3$	29.511
Fe(acac)$_3$	2.781
Fe(tfac)$_3$	8.340
Fe(hfac)$_3$	25.021

Table 8.5: Calculated vapor pressures, at 150 °C, for metal β-diketonate compounds (Figure 8.12) determined from thermogravimetric analysis. Data from B. D. Fahlman and A. R. Barron, Substituent effects on the volatility of metal β-diketonates. *Adv. Mater. Optics Electron.,* **2000, 10, 223.**

Quantification of impurities

Single walled carbon nanotubes (SWNTs, Figure 8.14) are typically synthesized using metal catalysts. Those prepared using the HiPco method, contain residual Fe catalyst (Figure 8.15). The metal (i.e., Fe) is usually oxidized upon exposure to air to the appropriate oxide (i.e., Fe_2O_3). While it is sometimes unimportant that traces of metal oxide are present during subsequent applications it is often necessary to quantify their presence. This is particularly true if the SWCNTs are to be used for cell studies since it has been shown that the catalyst residue is often responsible for observed cellular toxicity.

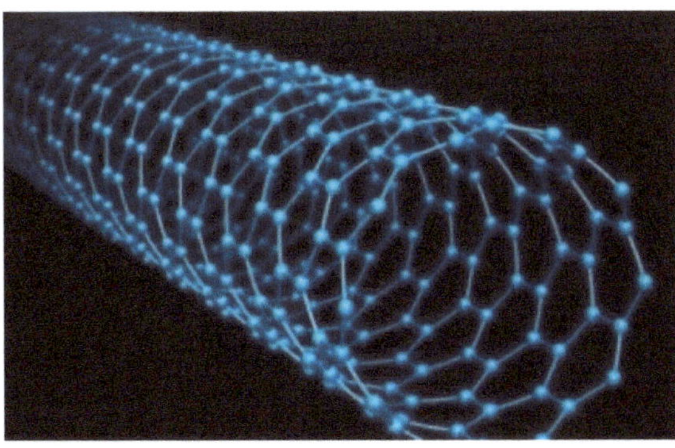

Figure 8.14: An example of the structure of a single walled carbon nanotube (SWCNT).

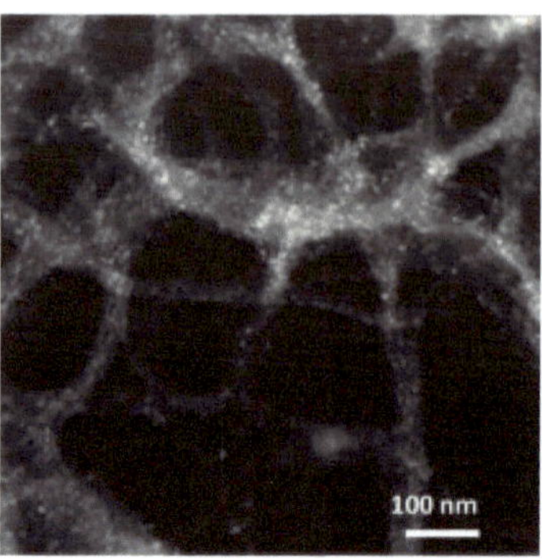

Figure 8.15: TEM micrographs of raw SWCNTs showing the residual catalyst (iron-containing) impurities. Adapted from V. Gomez, S. Irusta, W. W. Adams, R. H. Hauge, C. W. Dunnill, and A. R. Barron, Enhanced carbon nanotubes purification by physic-chemical treatment with microwave and Cl₂. *RSC Adv.*, 2016, 6, 11895. Copyright: Royal Society of Chemistry (2016).

In order to calculate the mass of catalyst residue the SWCNTs are pyrolyzed under air or O_2, and the residue is assumed to be the oxide of the metal catalyst. Water can be added to the raw SWCNTs, which enhances the low-temperature catalytic oxidation of carbon. A typical TGA plot of a sample of raw HiPco SWCNTs is shown in Figure 8.16.

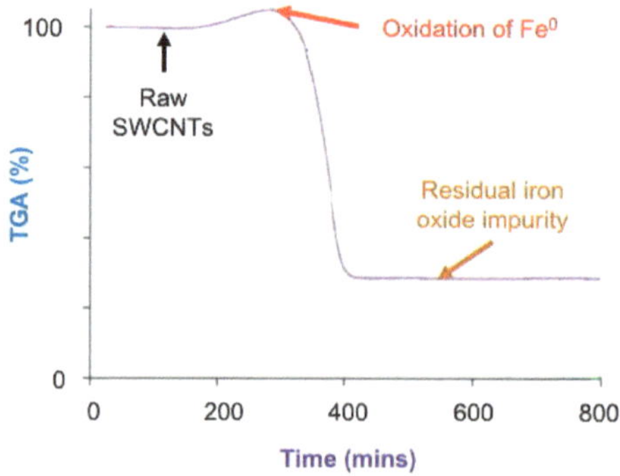

Figure 8.16: The TGA of unpurified HiPco SWNTs under air showing the residual mass associated with the iron catalyst. Adapted from I. W. Chiang, B. E. Brinson, A. Y. Huang, P. A. Willis, M. J. Bronikowski, J. L. Margrave, R. E. Smalley, and R. H. Hauge, Purification and characterization of single-wall carbon nanotubes (SWNTs) obtained from the gas-phase decomposition of CO (HiPco process). *J. Phys. Chem. B*, 2001, 105, 8297. Copyright: American Chemical Society (2001).

The weight gain (of ca. 5%) at 300 °C is due to the formation of metal oxide from the incompletely oxidized (Fe^0) catalyst. To determine the mass of iron catalyst impurity in the SWNT, the residual mass must be calculated. The residual mass is the mass that is left in the sample pan at the end of the experiment. From this TGA diagram, it is seen that 70% of the total mass is lost at 400 °C. This mass loss is attributed to the removal of carbon. The residual mass is 30%. Given that this is due to both oxide and oxidized metal, the original total mass of residual catalyst in raw HiPCO SWCNTs is ca. 25%.

Quantification of functional groups

Determining the number of functional groups on SWCNTs

The limitation of using SWCNTs in any practical applications is their solubility; for example, SWCNTs have little to no solubility in most solvents due to aggregation of the tubes. Aggregation/roping of nanotubes occurs as a result of the high van der Waals binding energy of ca. 500 eV per μm of tube contact. The van der Waals force between the tubes is so great, that it takes tremendous energy to pry them apart, making it very difficult to make combination of nanotubes with other materials such as in composite applications. The functionalization of nanotubes, i.e., the attachment of chemical functional groups, provides the path to overcome these barriers. Functionalization can improve solubility as well as processability and has been used to align the properties of nanotubes to those of other materials. In this regard, covalent functionalization provides a higher degree of fine-tuning for the chemical and physical properties of SWCNTs than non-covalent functionalization.

Functionalized nanotubes can be characterized by a variety of techniques, such as atomic force microscopy (AFM), transmission electron microscopy (TEM), UV-vis spectroscopy, and Raman spectroscopy, however, the quantification of the extent of functionalization is important and can be determined using TGA. Because any sample of functionalized-SWCNTs will have individual tubes of different lengths (and diameters) it is impossible to determine the number of substituents per SWCNT. Instead the extent of functionalization is expressed as number of substituents per SWNT carbon atom (C_{SWCNT}), or more often as C_{SWCNT}/substituent, since this is then represented as a number greater than 1.

Figure 8.17 shows a typical TGA for a functionalized SWNT. In this case it is polyethyleneimine (PEI, Figure 8.18) functionalized SWCNTs prepared by the reaction of fluorinated SWNTs (F-SWNTs) with PEI in the presence of a base catalyst.

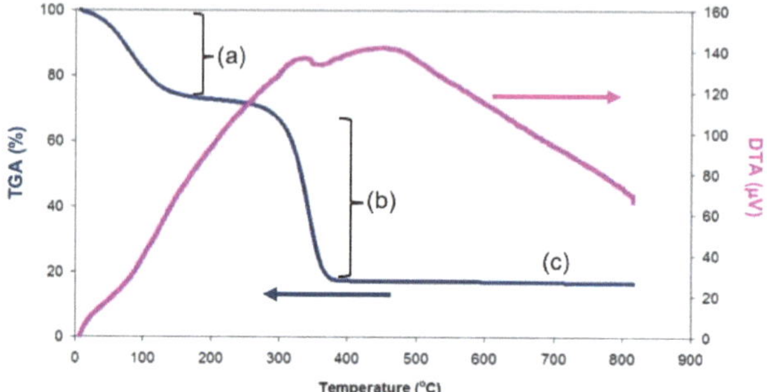

Figure 8.17: The TGA of SWNTs functionalized with polyethyleimine (PEI) under air showing the sequential loss of (a) complexed CO_2 and (b) decomposition of PEI, resulting in the formation of (c) unfunctionalized SWCNTs.

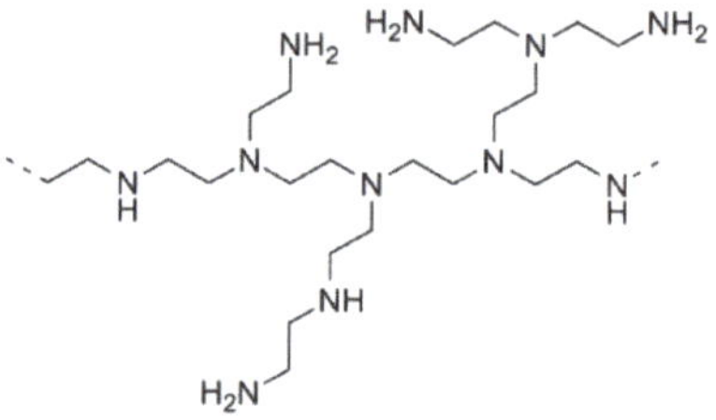

Figure 8.18: Structure of branched polyethyleneimine (PEI).

In the present case the molecular weight of the PEI is 600 g/mol. When the sample is heated, the PEI thermally decomposes leaving behind the unfunctionalized SWCNTs. The initial mass loss below 100 °C is due to residual water and ethanol used to wash the sample.

In the following example the total mass of the sample is 25 mg.
1. The initial mass, M_i = 25 mg = mass of the SWCNTs, residues and the PEI.
2. After the initial moisture has evaporated there is 68% of the sample left. 68% of 25 mg is 17 mg. This is the mass of the PEI and the SWCNTs.
3. At 300 °C the PEI starts to decompose and all of the PEI has been removed from the SWCNTs at 370 °C. The mass loss during this time is 53% of the total mass of the sample. 53% of 25 mg is 13.25 mg.

4. The molecular weight of this PEI is 600 g/mol. Therefore, there is 0.013 g/600 g/mol = 0.022 mmole of PEI in the sample.

5. 15% of the sample is the residual mass, this is the mass of the decomposed SWCNTs. Thus, 15% of 25 mg is 3.75 mg. The molecular weight of carbon is 12 g/mol. So, there is 0.3125 mmole of carbon in the sample.

6. There is 93.4 mol% of carbon and 6.5 mol% of PEI in the sample.

Determining the number of functional groups on metal oxide nanoparticles

Nanoparticles can be functionalized with wide array of polymers to tune their properties. For examples, in coating application, nanoparticles' surfaces are coated with hydrophobic ligands to prevent moisture migration to the metallic pipeline, extending the lifetime of the pipe. The same nanoparticles can be used for drug delivery application by using hydrophilic ligands to make the nanoparticles water soluble and biocompatible. Usually, nanoparticles focalization is done through ligand exchange, to get nanoparticles that are surrounded by organic polymers as shown in Figure 8.19. TGA is often employed in determination of surface coverage of organic ligands via heating the nanoparticles in air, decomposing the organic ligands without affecting the inorganic nanoparticle. Determination of surface coverage via TGA has an advantage over other microscopy techniques, such as transmission electron microscopy (TEM) because it is a straightforward method that requires no statistical refinement to get accurate results.

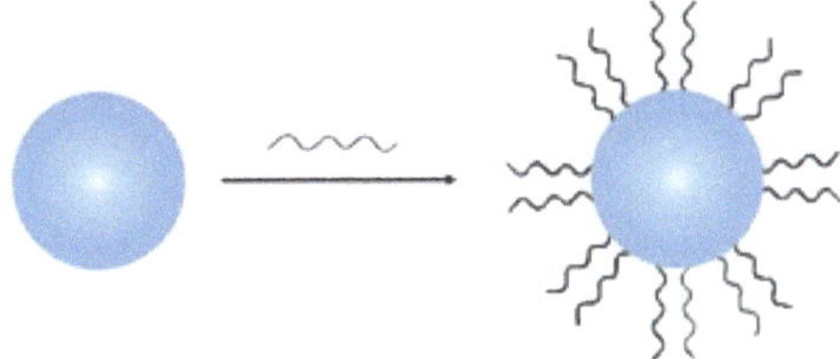

Figure 8.19: Illustration of nanoparticles functionalization process. Adapted from E. Mansfield, K. Tyner, C. Poling, and J. Blacklock, Determination of nanoparticle surface coatings and nanoparticle purity using microscale thermogravimetric analysis. *Anal. Chem.*, 2014, 7, 1478. Copyright: American Chemical Society (2014).

Data collection method

In the TGA software make sure to set the following parameters before starting the experimental procedure:

- Gas Selection. Select the experiment gas to be air. Make sure that the flow rate is around 40 mL/min for the balance and 60 mL/min for the furnace.
- Setting the heating program. Set the TGA to heat the sample from room temperature to 100 °C at a rate of 10 °C/min. Hold the temperature at 100 °C for 30 min. Continue heating at a rate 10 °C per min up to 800 °C.

Data analysis

After analysis use the TGA software to calculate the following:

- Weight loss from room temperature to 100 °C (Figure 8.20). This weight loss is attributed to the present of solvents/ moisture in the sample.
- Weight loss from 100 - 800 °C (Figure 8.20). This weight loss is attributed to decomposition of organic ligand on the surface of nanoparticles.

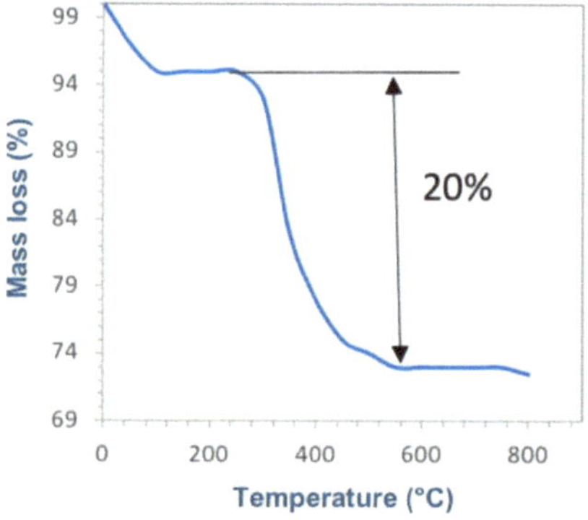

Figure 8.20: A typical TGA profile of functionalized nanoparticles. Adapted from E. Mansfield, K. Tyner, C. Poling, and J. Blacklock, Determination of nanoparticle surface coatings and nanoparticle purity using microscale thermogravimetric analysis. *Anal. Chem.*, 2014, 7, 1478. Copyright: American Chemical Society (2014).

Calculating surface coverage of the ligands on nanoparticles

1. Calculate the average particle diameter and average surface area of one nanoparticle using TEM. The volume of one particle (nm^3/particle) should be multiplied by the density (g/nm^3) of the nanoparticles, then, divide by the surface area of one particle (nm^2/particle). The final number is the inverse of the surface area of the nanoparticles in g/nm^2.
2. The number of polymer chains in 1 gram of nanoparticles should be determined by dividing the weight loss (g) by the molecular weight (g/mole) of the polymer and multiply that by Avogadro's number (chains/moles), then, divided by the weight at zero weight loss (g). The final number should have units of chains/gram.
3. Multiply the number from step 1 by the number from step 2 to get total surface coverage in chains/nm^2.

Adsorption processes

Solid-state ^{13}C NMR of PEI-SWNTs (c.f., Figure 8.18) shows the presence of carboxylate substituents that can be attributed to carbamate formation as a consequence of the reversable CO_2 absorption to the primary amine substituents of the PEI. Desorption of CO_2 is accomplished by heating under argon at 75 °C.

The quantity of CO_2 absorbed per PEI-SWNT unit may be determined by initially exposing the PEI-SWNT to a CO_2 atmosphere to maximize absorption. The gas flow is switched to either Ar or N_2 and the sample heated to liberate the absorbed CO_2 without decomposing the PEI or the SWNTs. An example of the appropriate TGA plot is shown in Figure 8.21.

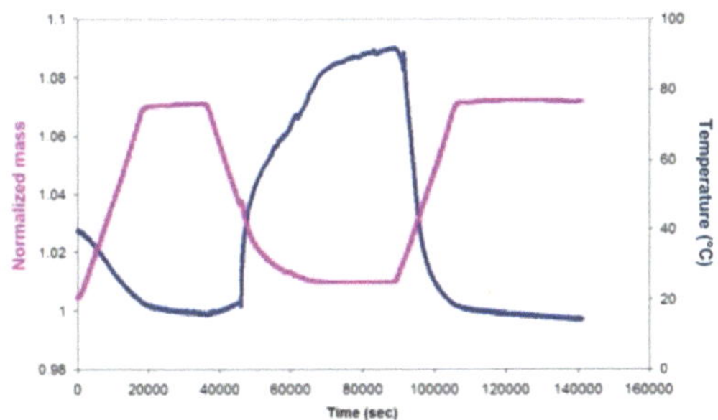

Figure 8.21: The TGA results of PEI(10000)-SWNT absorbing and desorbing CO_2. The mass has been normalized to the lowest mass recorded, which is equivalent to PEI(10000)-SWNT.

The sample was heated to 75 °C under Ar, and an initial mass loss due to moisture and/or atmospherically absorbed CO_2 is seen. In the temperature range of 25 - 75 °C the flow gas was switched from an inert gas to CO_2. In this region an increase in mass is seen, the increase is due to CO_2 absorption by the PEI (10000Da)-SWNT. Switching the carrier gas back to Ar resulted in the desorption of the CO_2.

The total normalized mass of CO_2 absorbed by the PEI(10000)-SWNT can be calculated as follows;

Solution outline

1. Minimum mass = mass of absorbant = $M_{absorbant}$
2. Maximum mass = mass of absorbant and absorbed species = M_{total}
3. Absorbed mass = $M_{absorbed}$ = M_{total} - $M_{absorbant}$
4. % of absorbed species= $(M_{absorbed}/M_{absorbant}) \times 100$
5. 1 mole of absorbed species = M_W of absorbed species
6. Number of moles of absorbed species = ($M_{absorbed}$ /M_W of absorbed species)
7. The number of moles of absorbed species absorbed per gram of absorbant= $(1\ g/M_{total}) \times$(Number of moles of absorbed species)

Solution

1. $M_{absorbant}$ = Mass of PEI-SWNT = 4.829 mg
2. M_{total} = Mass of PEI-SWNT and CO_2 = 5.258 mg
3. $M_{absorbed}$ = M_{total} - $M_{absorbant}$ = 5.258 mg - 4.829 mg = 0.429 mg
4. % of absorbed species= % of CO_2 absorbed = $(M_{absorbed}/M_{absorbant})$ $\times 100 = (0.429/4.829) \times 100 = 8.8\%$
5. 1 mole of absorbed species = M_W of absorbed species = M_W of CO_2 = 44 therefore 1 mole = 44g
6. Number of moles of absorbed species = ($M_{absorbed}/M_W$ of absorbed species) = (0.429 mg / 44 g) = 9.75 μM.
7. The number of moles of absorbed species absorbed per gram of absorbant = $(1\ g/M_{total}) \times$(Number of moles of absorbed species) = (1 g/5.258 mg)$\times$(9.75) = 1.85 mmol of CO_2 absorbed per gram of absorbant

High temperature TGA

Since the 1960's much work has gone into the development of new materials that could be used in harsh environments, especially for the increase of efficiency of vehicles that re-enter the Earth's atmosphere. The first systems that were studied were diborides with the goal of using these materials as a thermal protection system. In addition to acting as a protectant these ceramics also have the potential to allow for more sophisticated design from the traditional blunt capsules (rounded) that are present on current missiles and space shuttles. The benefit to these new designs if fabrication with UHTCs is implemented is that these vehicles will have more ability to be reused for further space explorations while also having the potential to explore farther with new paths of travel in space.

One of the most commonly considered UHTCs is ZrB_2 (Figure 8.22). This is the most promising contender for the application component of high temperature ceramics for two main reasons. First, this compound has a theoretical density of 6.09 g/cm^3, which is considered to be the lowest of all the UHTCs. This property makes it the most light and useful for aerospace application. In addition to this, ZrB_2 has a relatively high thermal conductivity (65 - 135 W/mK), allowing for optimal thermal shock resistance. Downsides to these materials is the fact that they begin to oxidize further at varying high temperatures, so this calls for extensive experimentation of the behavior of these UHTCs at high temperatures.

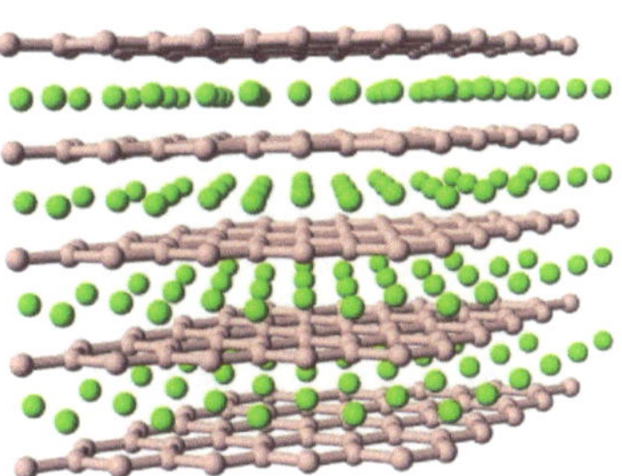

Figure 8.22: Structure of zirconium diboride (ZrB_2).

Analysis of the oxidation behavior of ZrB_2-SiC

One of the ceramic compounds that are of interest in ultra-high temperature applications is zirconium diboride. It has been proposed that at temperatures of 1500 °C ZrB_2-SiC would oxidize and produce a four-layer structure Figure

8.23. Based on these reports and the interest in further fundamental under-standing of the layer development, studies focused on experimentally describing the layers that develop when this process occurs. More specifi-cally, this case study considered ZrB_2 containing 30% vol SiC compound and its oxidation behavior when heated to 1500 °C in air.

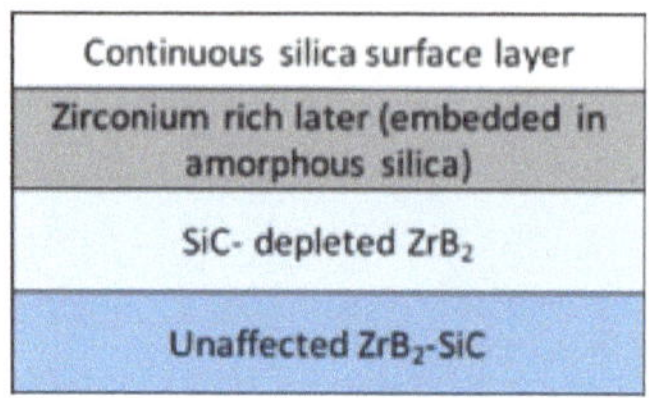

Figure 8.23: Reported result of oxidized ZrB$_2$-SiC compound following heating to 1500 °C in an air environment.

Prior to any experimentation, a ZrB_2-SiC blended powder was hot pressed and heated to result in a ZrB_2-SiC (30% volume) pellet, different pellets were used for all heating and TGA experiments. Two experiments were considered parallel to one another to analyze what was occurring in this compound. In a furnace (without TGA), a sample was heated to 1500 °C at a rate of 5 °C/ min. As for the TGA experiment, the weight change was measured under flowing air at a ramp rate of 5 °C/min up to 1500 °C (Figure 8.24). To determine the source of the mass change that was observed during the TGA experiment, experiment one's sample was characterized with scanning electron micros-copy (SEM) and energy dispersion spectroscopy (EDS).

Although not visible in SEM images of the samples, TGA did indicate a subtle change in mass that has been attributed to small amounts of ZrB_2-SiC oxida-tion with a rate of less than 0.2 mg/cm^2. As heating progressed (1000 °C) the weight gain that is present in Figure 8.24 is an indication of the beginning process of B_2O_3 and ZrO_2 formation. Figure 8.25a is a SEM of the formation of these oxide layers where the B_2O_3 layer (top layer) is approximately 2 mm thick and the ZrO_2-SiC (2nd layer) layer is ca. 6 mm thick. Due to the slower SiC oxidation, unoxidized particles are embedded in the zirconium oxide layer. However, when heated above 1200 °C the sample's layers are much different from what is apparent at 1000 °C. As shown in Figure 8.25b, the boron oxide layer that once existed has now evaporated off. This can also be confirmed with TGA Figure 8.24 because of the quick mass decrease and then continued increase. Through SEM and TGA combination this is the presence of SiO_2 formation, due to the slow beginning of SiC oxidation. Even further, as the temperatures approaches 1400 °C the SiC oxidizes even further, resulting in a larger SiO_2 layer (Figure 8.25c). This layer of SiO_2 acts as an oxidation protection layer to the original ZrB_2-SiC compound much more efficiently than the B_2O_3 layer that depleted after 1000 °C. Lastly, the final heating temperature that is considered in SEM (not TGA because it exceeds instrument limits) is 1500 °C. At this temperature the SiO_2 is not as protective as it appeared below such temperature. This is determined from the depleted SiO_2 and the addition of different ZrB_2-SiC byproducts (Figure 8.26).

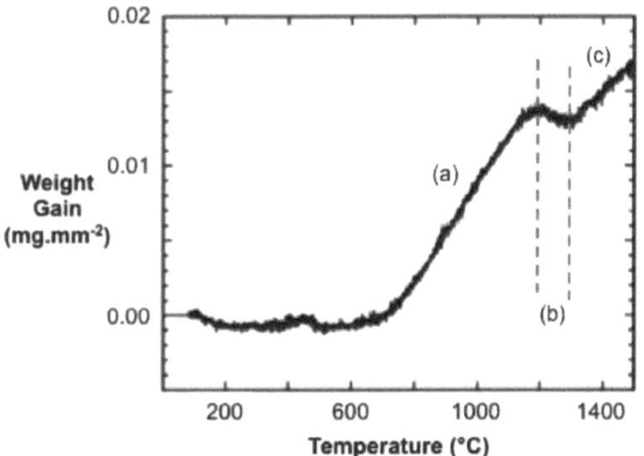

Figure 8.24: Thermal gravimetric analysis of 70-30 ZrB$_2$-SiC compound up to 1500 °C in air showing three distinct processes steps: (a) oxidation, (b) B$_2$O$_3$ sublimation, and (c) oxidation. Adapted from A. Rezaie, W. G. Fahrenholtz, and G. E. C. Hilmas, Evolution of structure during the oxidation of zirconium diboride–silicon carbide in air up to 1500 °C. *J. Eur. Ceram. Soc.*, 2007, 27, 2495. Copyright: Elsevier (2007).

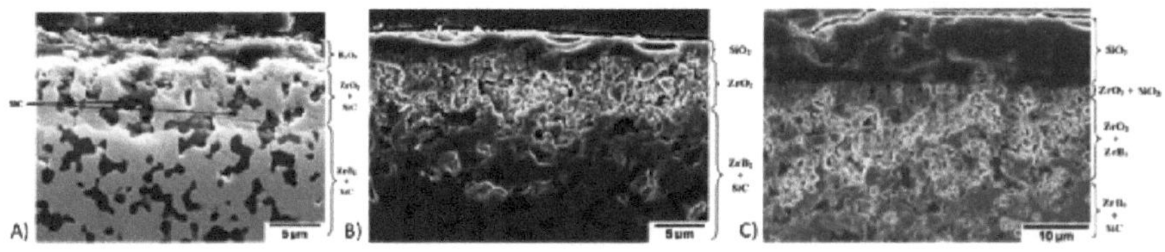

Figure 8.25: SEM images of 70-30 ratios of ZrB$_2$-SiC compounds at various temperatures to corroborate with the TGA analysis from Figure 2.104. (a) 1000 °C, (b) 1200 °C, and (c) 1400 °C. Adapted from A. Rezaie, W. G. Fahrenholtz, and G. E. C. Hilmas, Evolution of structure during the oxidation of zirconium diboride–silicon carbide in air up to 1500 °C. *J. Eur. Ceram. Soc.*, 2007, 27, 2495. Copyright: Elsevier (2007).

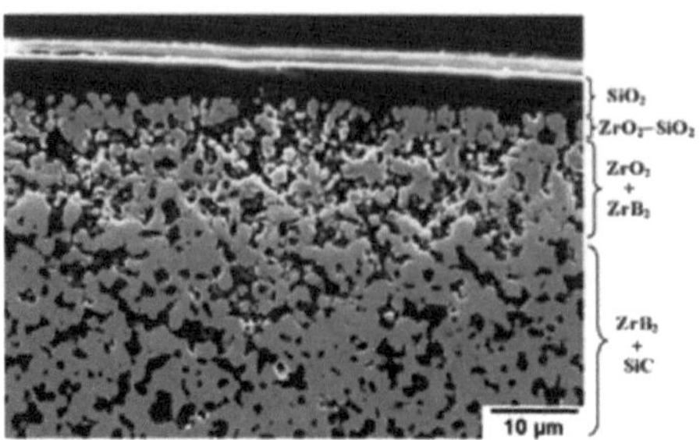

Figure 8.26: SEM image of ZrB₂-SiC pellet after heating to 1500 °C. Adapted from A. Rezaie, W. G. Fahrenholtz, and G. E. C. Hilmas, Evolution of structure during the oxidation of zirconium diboride–silicon carbide in air up to 1500 °C. *J. Eur. Ceram. Soc.*, 2007, 27, 2495. Copyright: Elsevier (2007).

High-temp isothermal oxidation of ULTCs

Since case study one, more work has been done to further understand the evolving oxidant behavior of UHTCs upon heating. TGA has multiple methods that can be used to consider the evolution of UHTCs. As a continuation to previous work that has been performed to better understand the mass changes that occur during the oxidation process of ZrB_2-SiC, this case study reports mass changes on oxidizing monolithic ZrB_2 + 25% vol SiC samples for short times at temperatures above 1000 °C. The first technique used is the conventional method (as seen in case study one), where the sample is placed in a TGA furnace and mass change is measured as a function of temperature with a heating rate of 20 °C/min while being exposed to air flow. In this method during any isothermal hold there is the potential for heating and cooling shoulder as reported in many other conventional heating methods.

Contrary to the conventional TGA method, there are two other techniques that can be utilized to examine this evolution of UHTCs with the absence of heating and cooling shoulders. One means of analysis is quick insertion of the sample at the temperature of interest. This method's complication arises in the measuring of the mass loss because periodic removal of the sample, cooling time, mass recording, and then reinsertion at next desired temperature is required. The second method focuses on heating the material to the desired temperature in the presence of an inert gas and once the temperature is reached introducing the reactive gas. This case study chooses to focus on the second of these two methods for convenience and more control of the experiment.

This method will now be referred to as dynamic non-equilibrium TGA (DNE). The schematic shown in Figure 8.27 is a visual of the set up that is used for the DNE technique. The final technique that was used for this set of experiments was a conventional and DNE combination.

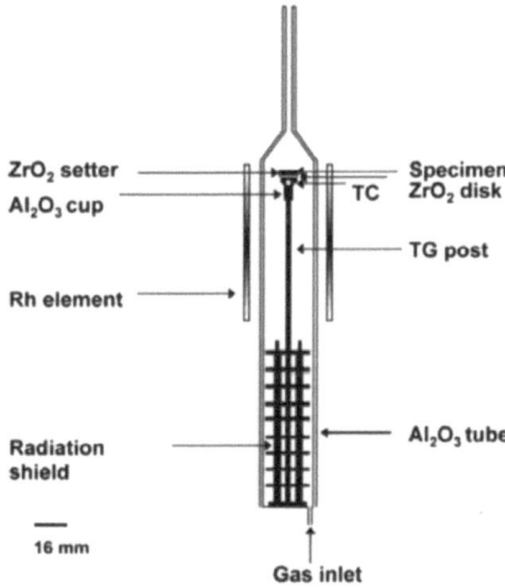

Figure 8.27: Experimental set-up for non-conventional technique of interest when reaction gas is implemented onto the sample instantaneously. Adapted from M. Miller-Oana and E. L. Corral, High-temperature isothermal oxidation of ultra-high temperature ceramics using thermal gravimetric analysis. *J. Am. Ceram. Soc.*, 2016, 99, 619. Copyright: American Chemical Society (2016).

As for the specifics in sample testing, these compounds were considered at short times (<15 minutes) at high temperatures because the main focus was to better understand how quickly oxide layers form. In the DNE-TGA experiment, the sample was heated 20 °C/ min in pure argon gas (100 mL/min flow rate) to the isothermal temperature (1600 °C). Following a two-minute equilibrium period air was released into the set-up. The instrument sat with air in its chamber for 15 minutes and then the system was purged with argon again at a rate of 50 mL/min and is followed by cooling.

To compare the DNE method to the conventional technique another identical sample was heated to 1600 °C at 20 °C/min and settled at this temperature for 15 minutes. Lastly, to compare these methods closely together, a combination method (described earlier) was performed. In this, the sample was first placed in the conventional furnace where it was heated to 1600 °C while being exposed to air. The sample was then cooled to 500 °C, switched to argon gas and then heated to 1600 °C under the inert gas and then exposed to air for 15 minutes at this stable temperature.

From this experiment Figure 8.28 was obtained. From which it can be determined that when comparing the DNE method to the combo technique the %mass change is different. This is because in the combination method the slow heating that occurs at the beginning of this experiments allows for the slow formation of an oxide layer that acts as a protectant to further quick oxidation. Contrary to this date, the DNE method has a steep increase in mass change. This is because of the instant exposure to air which prohibits a slow oxide layer formation, rather, this technique results in the quick formation of a thicker oxide layer.

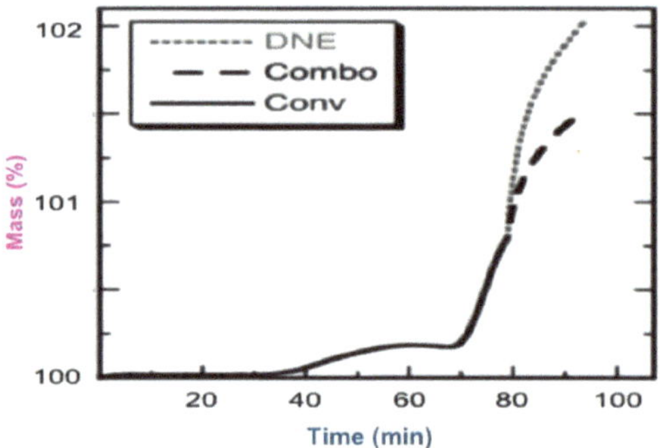

Figure 8.28: Weight variation of dynamic non-equilibrium (DNE), combination, and conventional TGA experiments on ZrB$_2$-SiC at 1600 °C. Adapted from M. Miller-Oana and E. L. Corral, High-temperature isothermal oxidation of ultra-high temperature ceramics using thermal gravimetric analysis. *J. Am. Ceram. Soc.*, 2016, 99, 619. Copyright: American Chemical Society (2016).

From this comparison experiment where different environment exposures (air and argon) were considered at a similar heating rate and hold time, it can be concluded that this new TGA method indicates that more mass is gained when compared to the conventional method. This is also prevalent with the use of SEM following the heating experiments. In the conventional TGA sample there is a glassy SiO$_2$ layer, whereas the DNE method does have this layer, but additionally contains boron oxides which results in a less protectant layer. From the entirety of this case study, it was also concluded that ZrB$_2$ + SiC oxidation begins at ~1300 °C.

TGA–FTIR analysis

Fourier transform infrared spectroscopy

Fourier transform infrared spectroscopy (FTIR) is the most popular spectroscopic method used for characterizing organic and inorganic compounds. The basic modification of an FTIR from a regular IR instrument is a device called interferometer, which generates a signal that allows very fast IR spectrum acquisition. For doing so, the generated interferogram has to be expanded using a Fourier transformation to generate a complete IR frequency spectrum. In the case of performing FTIR transmission studies the intensity of the transmitted signal is measured, and the IR fingerprint is generated,

$$T = I/I_b = e^{c\varepsilon l}$$

where I is the intensity of the samples, I_b is the intensity of the background, c is the concentration of the compound, ε is the molar extinction coefficient and l is the distance that light travels through the material. A transformation of transmission to absorption spectra is usually performed and the actual concentration of the component can be calculated by applying the Beer-Lambert law,

$$A = -\ln(T) = c\varepsilon l$$

A qualitative IR-band map is presented in Figure 8.29. The absorption bands between 4000 - 1600 cm^{-1} represent the group frequency region and are used to identify the stretching vibrations of different bonds. At lower frequencies (from 1600 - 400 cm^{-1}) vibrations due to intermolecular bond bending occurs upon IR excitation and therefore are usually not taken into account.

TGA-FTIR characterization

TGA is a powerful tool for identifying the different compounds evolved during the controlled pyrolysis and therefore provide qualitative and quantitative information about the volatile components of the sample. In metal oxide nanoparticle synthesis TGA-FTIR studies can provide qualitative and quantitative information about the volatile compounds of the nanoparticles.

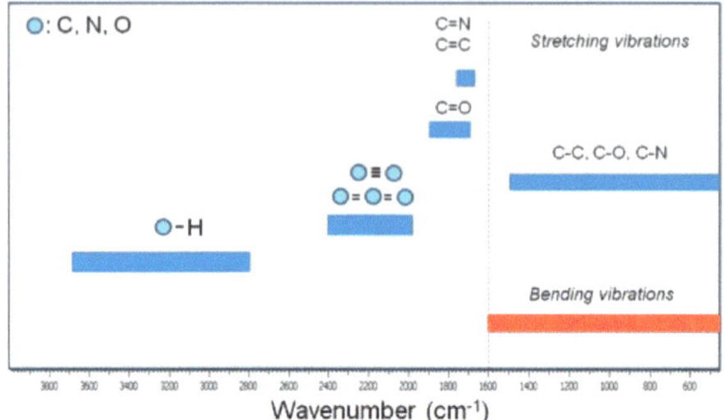

Figure 8.29: Selected FTIR stretching and bending modes associated with the typical ligands used for nanoparticle stabilization.

WO_{3-x} nanorods (Figure 8.30) were prepared from a mixture of trimethylamine N-oxide (TMAO, Figure 8.31), oleylamine (Figure 8.32), and $W(CO)_6$ (Figure 8.33) with heating to 250 °C at a rate of 3 °C/min. The TGA/DSC plot of the as prepared WO_{3-x} nanorods is shown in Figure 8.34.

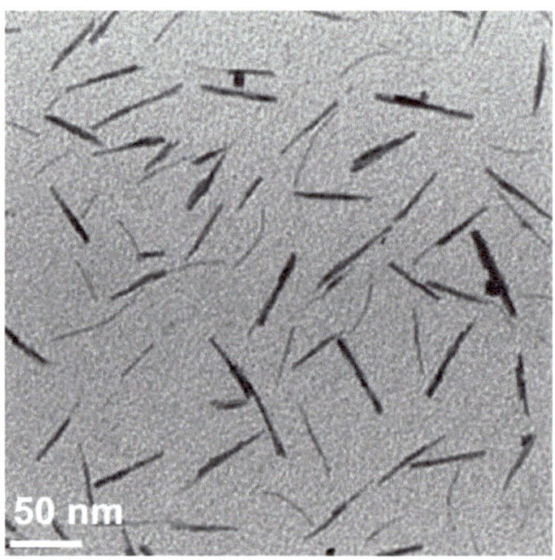

Figure 8.30: TEM image of WO_{3-x} nanorods

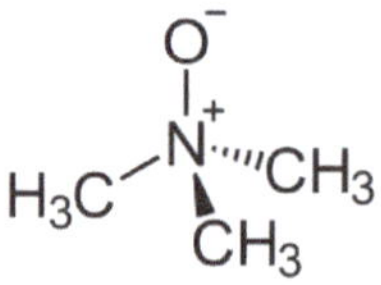

Figure 8.31: Structure of trimethylamine N-oxide (Me₃NO, TMAO).

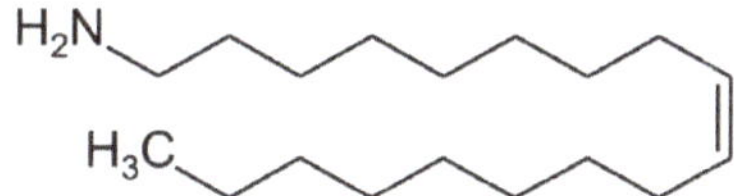

Figure 8.32: Structure of oleylamine ($C_{18}H_{35}NH_2$).

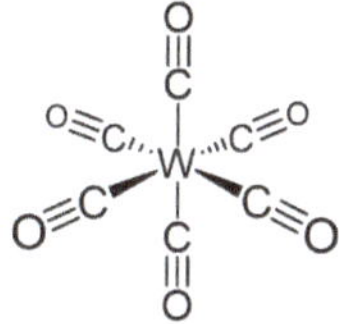

Figure 8.33: Tungsten hexacarbonyl, $W(CO)_6$.

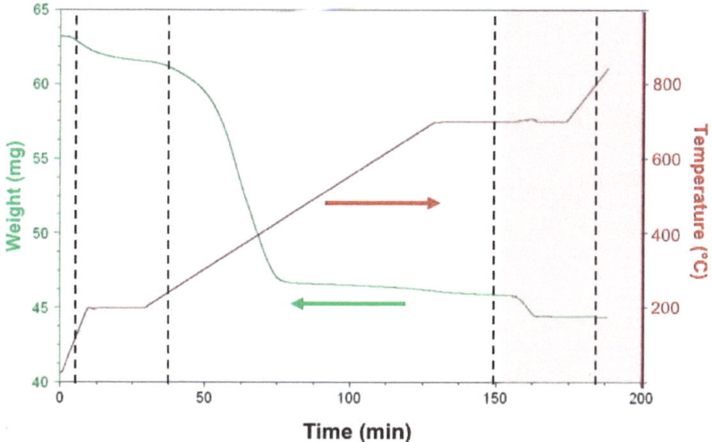

Figure 8.34: TGA of WO_{3-x} nanorods (green line) showing the experimental temperature profile (red line). The shaded region is where the carrier gas was switched to air. The dashed lines represent the time for the FT-IR spectra plotted in Figure 8.35.

Selected IR spectra are presented in Figure 8.35 are for the TGA after 4, 37, 149 and 185 minutes (Figure 8.34) for WO_{3-x} nanorods. Four regions with intense peaks are observed:

- 4000 - 3550 cm^{-1} due to O-H bond stretching assigned to H_2O that is always present and due to N-H group stretching that is assigned to the amine group of oleylamine (Figure 8.32).
- 2400 - 2250 cm^{-1} due to O=C=O stretching,

- 1900 - 1400 cm^{-1}, which is mainly to C=O stretching.
- 800 - 400 cm^{-1} cannot be resolved.

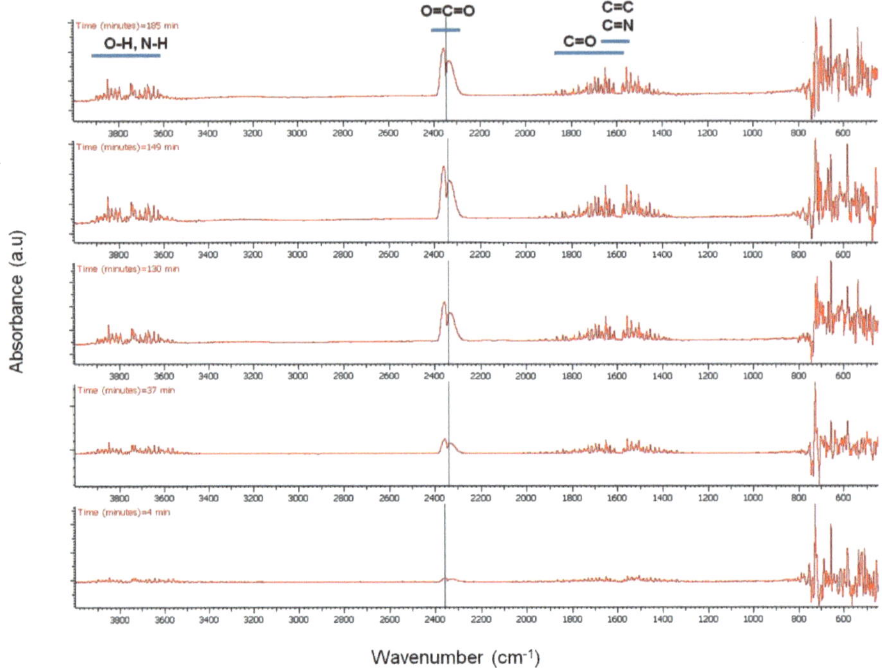

Figure 8.35: FTIR spectra of products from the TGA after (4, 37, 149 and 185 minutes) of WO$_{3-x}$ nanorods. The point that each IR spectra was collected is indicated by the dashed lines in Figure 8.33.

The peak intensity evolution with time can be more easily observed in Figure 8.36 and Figure 8.37. As seen, CO$_2$ evolution increases significantly with time especially after switching our flow from N$_2$ to air. H$_2$O seems to be present in the out flow stream up to 700 °C while the majority of the N-H amine peaks seem to disappear at about 75 min. C=N compounds are not expected to be present in the stream which leaves bands between 1900 1400 cm^{-1} assigned to C=C and C=O stretching vibrations. Unsaturated olefins resulting from the cracking of the oleylamine molecule are possible at elevated temperatures as well as the presence of CO especially under N$_2$ atmosphere.

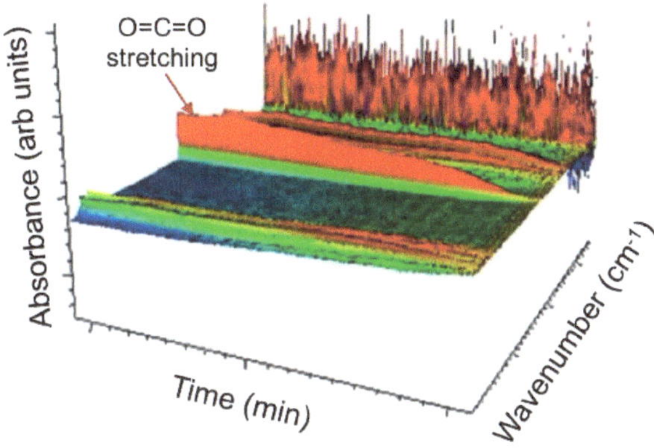

Figure 8.36: 3D representation of time evolution of the FT-IR spectra of the volatile compounds formed during the TGA of WO₃₋ₓ nanorods.

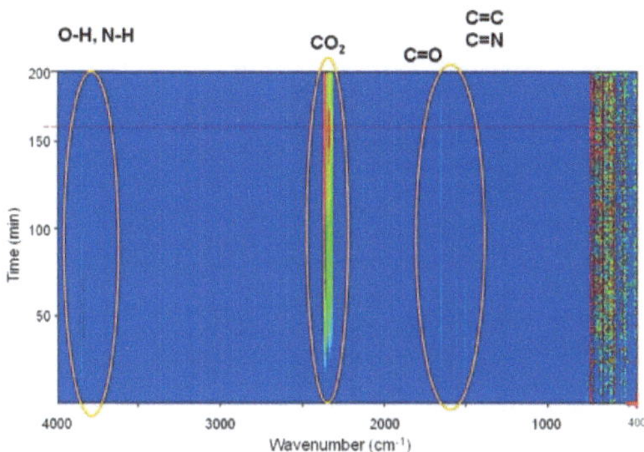

Figure 8.37: Intensity profile of FTIR spectra of the volatile compounds formed from the pyrolysis of WO₃₋ₓ nanorods.

From the above compound identification, we can summarize and propose the following applications for TGA-FTIR. First, more complex ligands, containing aromatic rings and maybe other functional groups may provide more insight in the ligand to MO_x interaction. Second, the presence of CO and CO_2 even under N_2 flow means that complete O_2 removal from the TGA and the FTIR cannot be achieved under these conditions. Even though the system was equilibrated for more than an hour, traces of O_2 are existent which create errors in our calculations.

Bibliography

P. W. Atkins, *Physical Chemistry*, 5th edn., W. H. Freeman, New York (1994).

G. Beech and R. M. Lintonbon, Thermal and kinetic studies of some complexes of 2,4-pentanedione. *Thermochim. Acta*, 1971, **3**, 97.

L. M. Bronstein, X. Huang, J. Retrum, A. Schmucker, M. Pink, B. D. Stein, and B. Dragnea, Influence of iron oleate complex structure on iron oxide nanoparticle formation. *Chem. Mater.*, 2007, **19**, 3624.

I. W. Chiang, B. E. Brinson, A. Y. Huang, P. A. Willis, M. J. Bronikowski, J. L. Margrave, R. E. Smalley, and R. H. Hauge, Purification and characterization of single-wall carbon nanotubes (SWNTs) obtained from the gas-phase decomposition of CO (HiPco process). *J. Phys. Chem. B*, 2001, **105**, 8297.

E. P. Dillon, C. A. Crouse and A. R. Barron, Synthesis, characterization, and carbon dioxide adsorption of covalently attached polyethyleneimine-functionalized single-wall carbon nanotubes. *ACS Nano*, 2008, **2**, 156.

B. D. Fahlman, PhD Thesis, *Chemical Vapor Deposition of Alumina-Based Thin Films*, Department of Chemistry, Rice University (2000).

B. D. Fahlman and A. R. Barron, Substituent effects on the volatility of metal β-diketonates. *Adv. Mater. Optics Electron.*, 2000, **10**, 223.

E. G. Gillan, S. G. Bott, and A. R. Barron, Volatility studies on gallium chalcogenide cubanes: thermal analysis and determination of sublimation enthalpies. *Chem. Mater.*, 1997, **9**, 796.

V. Gomez, S. Irusta, W. W. Adams, R. H. Hauge, C. W. Dunnill, and A. R. Barron, Enhanced carbon nanotubes purification by physic-chemical treatment with microwave and Cl_2. *RSC Adv.*, 2016, **6**, 11895.

G. R. Patzke, F. Krumeich, and R. Nesper, Oxidic nanotubes and nanorods—anisotropic modules for a future nanotechnology. *Angew. Chem., Int. Ed.*, 2002, **41**, 2446.

J. Han, P. Hu, X. Zhang, and S. Meng, Oxidation behavior of zirconium diboride–silicon carbide at 1800 °C. *Scr. Mater.*, 2007, **57**, 825.

J. Han, P. Hu, X. Zhang, S. Meng, and W. Han, Oxidation-resistant ZrB_2–SiC composites at 2200 °C. *Compos. Sci. Technol.*, 2008, **68**, 799.

J. O. Hill and J. P. Murray, The sublimation enthalpies of metal complexes of pentane-2,4-dione and related ligands - an assessment review. *Rev. Inorg. Chem.*, 1993, **13**, 125.

P. Hu, W. Guolin, and Z, Wang, Oxidation mechanism and resistance of ZrB_2-SiC composites. *Corros. Sci.*, 2009, 51, 2724.

K. Lee, W. S. Seo, and J. T. Park, Synthesis and optical properties of colloidal tungsten oxide nanorods. *J. Am. Chem. Soc.*, 2003, **125**, 3408.

E. Mansfield, K. Tyner, C. Poling, and J. Blacklock, Determination of nanoparticle surface coatings and nanoparticle purity using microscale thermogravimetric analysis. *Anal. Chem.*, 2014, **7**, 1478.

M. Miller-Oana and E. L. Corral, High-temperature isothermal oxidation of ultra-high temperature ceramics using thermal gravimetric analysis. *J. Am. Ceram. Soc.*, 2016, **99**, 619.

J. P. Murray, K. J. Cavell and J. O. Hill, DSC determination of the fusion and sublimation enthalpy of tris(2,4-pentanedionato)chromium(III) and iron(III). *Thermochim. Acta*, 1980, **36**, 97.

A. Rezaie, W. G. Fahrenholtz, and G. E. C. Hilmas, Evolution of structure during the oxidation of zirconium diboride–silicon carbide in air up to 1500 °C. *J. Eur. Ceram. Soc.*, 2007, **27**, 2495.

M. A. V. Ribeiro da Silva and M. L. C. C. H. Ferrao, Vapour pressures and standard molar enthalpies of sublimation of beryllium(II) and aluminium(III) benzoylacetonates. *J. Chem. Thermodyn.*, 1994, **26**, 315.

R. Sabbah, D. Tabet, S. Belaadi, Enthalpie de sublimation ou vaporisation de quelques dérivés méthylés du benzène. *Thermochim. Acta*, 1994, **247**, 193.

L. Tong, E. Lu, J. Pichaandi, P. Cao, M. Nitz, and M. Winnik, Quantification of surface ligands on $NaYF_4$ nanoparticles by three independent analytical techniques. *Chem. Mater.*, 2015, **27**, 4899.

L. A. Torres-Gomez, G. Barreiro-Rodriquez, and A. Galarza-Mondragon, A new method for the measurement of enthalpies of sublimation using differential scanning calorimetry. *Thermochim. Acta*, 1988, **124**, 229.

152

Chapter 9: Differential Scanning Calorimetry

Christopher Burbridge, Basil Shadfan, Pavan M. V. Raja
and Andrew R. Barron

Introduction

Differential scanning calorimetry (DSC) is a technique used to measure the difference in the heat flow rate of a sample and a reference over a controlled temperature range. These measurements are used to create phase diagrams and gather thermo-analytical information such as transition temperatures and enthalpies.

History

DSC was developed in 1962 by Perkin-Elmer employees Emmett Watson and Michael O'Neill (Figure 9.1) and was introduced at the Pittsburgh Conference on Analytical Chemistry and Applied Spectroscopy. The equipment for this technique was available to purchase beginning in 1963 and has evolved to control temperatures more accurately and take measurements more precisely, ensuring repeatability and high sensitivity.

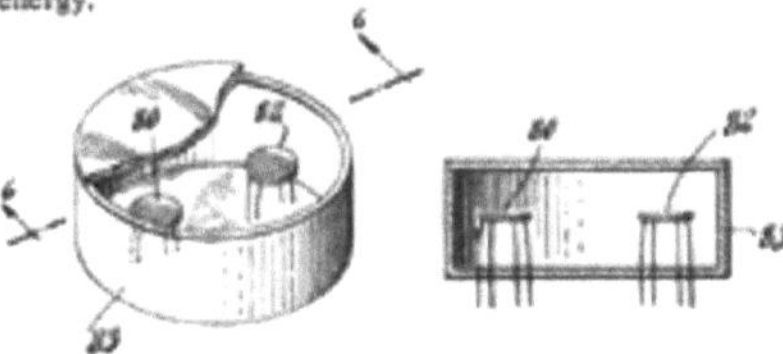

DIFFERENTIAL MICROCALORIMETER
Emmett S. Watson, Ridgefield, and Michael J. O'Neill, West Redding, Conn., assignors to The Perkin-Elmer Corporation, Norwalk, Conn., a corporation of New York

Filed Apr. 4, 1962, Ser. No. 185,499
24 Claims. (Cl. 73—15)

1. The method of performing an analysis which comprises varying the environment of a sample material; measuring the resulting difference in temperature between said sample material and a reference material; varying the relative flow of thermal energy between both said sample and said reference material relative to at least one external energy source in response to said difference in temperature in such manner as to equalize the temperature of said sample and said reference material; and independently varying an additional heat flow to both said sample and reference material in such manner as to cause them both to attain the same desired temperature; and measuring said first-mentioned relative flow of thermal energy.

Figure 9.1: An excerpt from the original US Patent for DSC.

Theory

Phase transitions

Phase transitions refer to the transformation from one state of matter to another. Solids, liquids, and gasses are changed to other states as the thermodynamic system is altered, thereby affecting the sample and its properties. Measuring these transitions and determining the properties of the sample is important in many industrial settings and can be used to ensure purity and determine composition (such as with polymer ratios). Phase diagrams (Figure 9.2) can be used to clearly demonstrate the transitions in graphical form, helping visualize the transition points and different states as the thermodynamic system is changed.

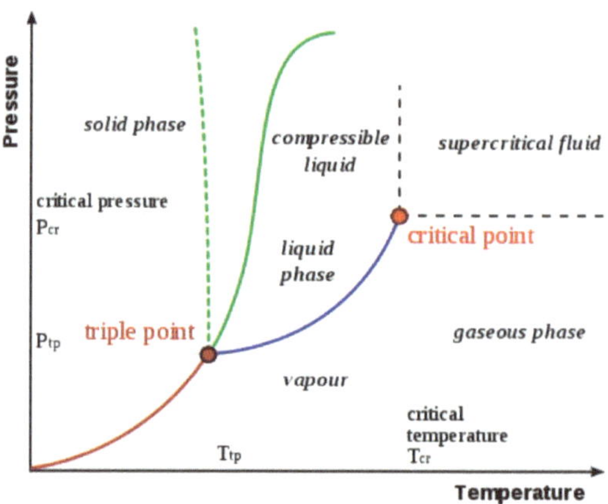

Figure 9.2: An example of a typical phase diagram.

Differential thermal analysis

Prior to DSC, differential thermal analysis (DTA) was used to gather information about transition states of materials. In DTA, the sample and reference are heated simultaneously with the same amount of heat and the temperature of each is monitored independently. The difference between the sample temperature and the reference temperature gives information about the exothermic or endothermic transition occurring in the sample. This strategy was used as the foundation for DSC, which sought to measure the difference in energy needed to keep the temperatures the same instead of measuring the difference in temperature from the same amount of energy.

Differential scanning calorimeter

Instead of measuring temperature changes as heat is applied as in DTA, DSC measures the amount of heat that is needed to increase the temperatures of the sample and reference across a temperature gradient. The sample and reference are kept at the same temperature as it changes across the gradient, and the differing amounts of heat required to keep the temperatures synchronized are measured. As the sample undergoes phase transitions, more or less heat is needed, which allows for phase diagrams to be created from the data. Additionally, specific heat, glass transition temperature, crystallization temperature, melting temperature, and oxidative/thermal stability, among other properties, can be measured using DSC.

Applications

DSC is often used in industrial manufacturing, ensuring sample purity and confirming compositional analysis. Also used in materials research, providing information about properties and composition of unknown materials can be determined. DSC has also been used in the food and pharmaceutical industries, providing characterization and enabling the fine-tuning of certain properties. The stability of proteins and folding/unfolding information can also be measured with DSC experiments.

Instrumentation

Equipment

The sample and reference cells (also known as pans), each enclosing their respective materials, are contained in an insulted adiabatic chamber (Figure 9.3). The cells can be made of a variety of materials, such as aluminum, copper, gold and platinum. The choice of which is dictated by the necessary upper temperature limit. A variable heating element around each cell transfers heat to the sample, causing both cells' temperature to rise in coordination with the other cell. A temperature monitor measures the temperatures of each cell and a microcontroller controls the variable heating elements and reports the differential power required for heating the sample versus the reference. A typical setup, including a computer for controlling software, is shown in Figure 9.4.

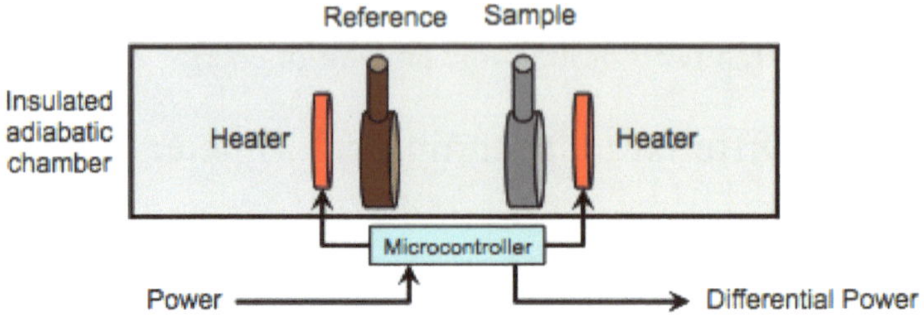

Figure 9.3: Diagram of basic DSC equipment.

Figure 9.4: Picture of a basic DSC setup in a laboratory.

Modes of operations

With advancement in DSC equipment, several different modes of operations now exist that enhance the applications of DSC. Scanning mode typically refers to conventional DSC, which uses a linear increase or decrease in temperature. An example of an additional mode often found in newer DSC equipment is an isothermal scan mode, which keeps temperature constant while the differential power is measured. This allows for stability studies at constant temperatures, particularly useful in shelf life studies for pharmaceutical drugs.

Calibration

As with practically all laboratory equipment, calibration is required. Calibration substances, typically pure metals such as indium or lead, are chosen that have clearly defined transition states to ensure that the measured transitions correlate to the literature values.

Obtaining measurements

Sample preparation

Sample preparation mostly consists of determining the optimal weight to analyze. There needs to be enough of the sample to accurately represent the material, but the change in heat flow should typically be between 0.1 - 10 mW. The sample should be kept as thin as possible and cover as much of the base of the cell as possible. It is typically better to cut a slice of the sample rather than crush it into a thin layer. The correct reference material also needs to be determined in order to obtain useful data.

DSC curves

DSC curves (e.g., Figure 9.5) typically consist of the heat flow plotted versus the temperature. These curves can be used to calculate the enthalpies of transitions, (ΔH), by integrating the peak of the state transition,

$$\Delta H = KA$$

where K is the calorimetric constant and A is the area under the curve.

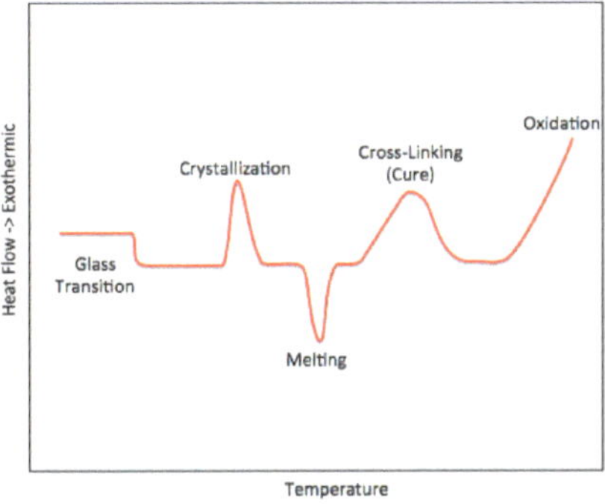

Figure 9.5: An idealized DSC curve showing the shapes associated with particular phase transitions.

Sources of error

Common error sources apply, including user and balance errors and improper calibration. Incorrect choice of reference material and improper quantity of sample are frequent errors. Additionally, contamination and how the sample is loaded into the cell affect the DSC.

DSC characterization of polymers

Differential scanning calorimetry (DSC), at the most fundamental level, is a thermal analysis technique used to track changes in the heat capacity of some substance. To identify this change in heat capacity, DSC measures heat flow as a function of temperature and time within a controlled atmosphere. The measurements provide a quantitative and qualitative look into the physical and chemical alterations of a substance related to endothermic or exothermic events.

Overview of polymeric properties

A polymer is, essentially, a chemical compound whose molecular structure is a composition of many monomer units bonded together (Figure 9.6). The physical properties of a polymer and, in turn, its thermal properties are determined by this very ordered arrangement of the various monomer units that compose a polymer. The ability to correctly and effectively interpret differential scanning calorimetry data for any one polymer stems from an understanding of a polymer's composition. As such, some of the more essential dynamics of polymers and their structures are briefly addressed below.

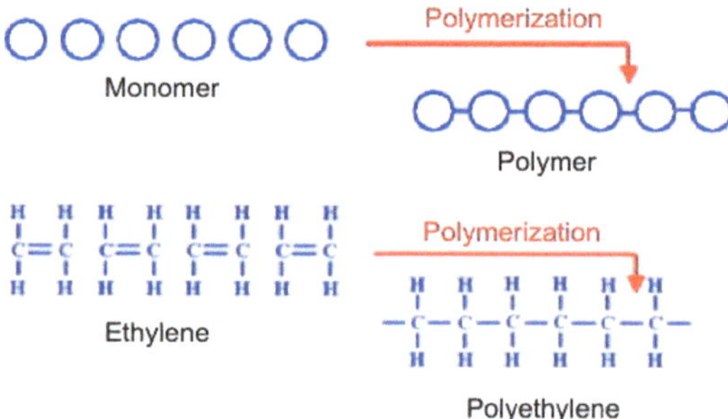

Figure 9.6: Schematic diagram of monomer circles polymerizing to form a polymer chain, and by analogy, ethylene monomers polymerizing to form polyethylene.

An aspect of the ordered arrangement of a polymer is its degree of polymerization, or, more simply, the number of repeating units within a polymer chain. This degree of polymerization plays a role in determining the molecular weight of the polymer. The molecular weight of the polymer, in turn, plays a role in determining various thermal properties of the polymer such as the perceived melting temperature.

Related to the degree of polymerization is a polymer's dispersity, i.e., the uniformity of size among the particles that compose a polymer. The more uniform a series of molecules, the more monodisperse the polymer; however, the more non-uniform a series of molecules, the more polydisperse the polymer. Increases in initial transition temperatures follow an increase in polydispersity. This increase is due to higher intermolecular forces and polymer flexibility in comparison to more uniform molecules.

In continuation with the study of a polymer's overall composition is the presence of cross-linking between chains. The ability for rotational motion within a polymer decreases as more chains become cross-linked, meaning initial transition temperatures will increase due to a greater level of energy needed to overcome this restriction. In turn, if a polymer is composed of stiff functional groups, such as carbonyl groups, the flexibility of the polymer will drastically decrease, leading to higher transitional temperatures as more energy will be required to break these bonds. The same is true if the backbone of a polymer is composed of stiff molecules, like aromatic rings, as this also causes the flexibility of the polymer to decrease. However, if the backbone or internal structure of the polymer is composed of flexible groups, such as aliphatic chains, then either the packing or flexibility of the polymer decreases. Thus, transitional temperatures will be lower as less energy is needed to break apart these more flexible polymers.

Lastly, the actual bond structure (i.e., single, double, triple) and chemical properties of the monomer units will affect the transitional temperatures. For examples, molecules more predisposed towards strong intermolecular forces, such as molecules with greater dipole to dipole interactions, will result in the need for higher transitional temperatures to provide enough energy to break these interactions.

In terms of the relationship between heat capacity and polymers: heat capacity is understood to be the amount of energy a unit or system can hold before its temperature raises one degree; further, in all polymers, there is an increase in heat capacity with an increase in temperature. This is due to the fact that as polymers are heated, the molecules of the polymer undergo greater levels of rotation and vibration which, in turn, contribute to an increase in the internal energy of the system and thus an increase in the heat capacity of the polymer.

In knowing the composition of a polymer, it becomes easier to not only pre-emptively hypothesize the results of any DSC analysis but also troubleshoot why DSC data does not seem to corroborate with the apparent properties of a polymer. Note, too, that there are many variations in DSC techniques and types as they relate to characterization of polymers. These differences are discussed below.

Standard DSC (heat flux DSC)

The composition of a prototypical, unmodified DSC includes two pans. One is an empty reference plate and the other contains the polymer sample. Within the DSC system is also a thermoelectric disk. Calorimetric measurements are then taken by heating both the sample and empty reference plate at a controlled rate, say 10 °C/min, through the thermoelectric disk. A purge gas is admitted through an orifice in the system, which is preheated by circulation through a heating block before entering the system. Thermocouples within the thermoelectric disk then register the temperature difference between the two plates. Once a temperature difference between the two plates is measured, the DSC system will alter the applied heat to one of the pans so as to keep the temperature between the two pans constant. In Figure 9.7 is a cross-section of a common heat flux DSC instrument.

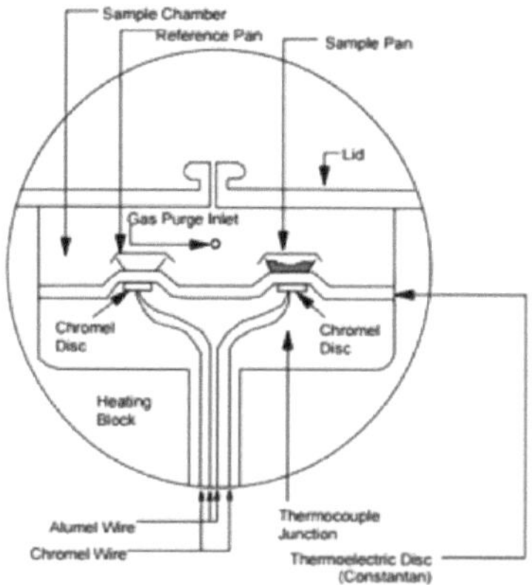

Figure 9.7: Schematic diagram of a heat flux DSC. Used with permission Copyright TA Instruments.

The resulting plot that is one in which the heat flow is understood to be a function of temperature and time. As such, the slope at any given point is proportional to the heat capacity of the sample. The plot as a whole, however, is representative of thermal events within the polymer. The orientation of peaks or stepwise movements within the plot, therefore, lend themselves to interpretation as thermal events.

To interpret these events, it is important to de ne the thermodynamic system of the DSC instrument. For most heat flux systems, the thermodynamic system is understood to be only the sample. This means that when, for example, an exothermic event occurs, heat from the polymer is released to the outside environment and a positive change is measured on the plot. As such, all exothermic events will be positive shifts within the plot while all endothermic events will be negative shifts within the plot. However, this can be flipped within the DSC system, so be sure to pay attention to the orientation of your plot as *exo* up or *exo* down. See Figure 9.8 for an example of a standard DSC plot of polymer poly(ethylene terephthalate) (PET, Figure 9.9). By understanding this relationship within the DSC system, the ability to interpret thermal events, such as the ones described below, becomes all the more approachable.

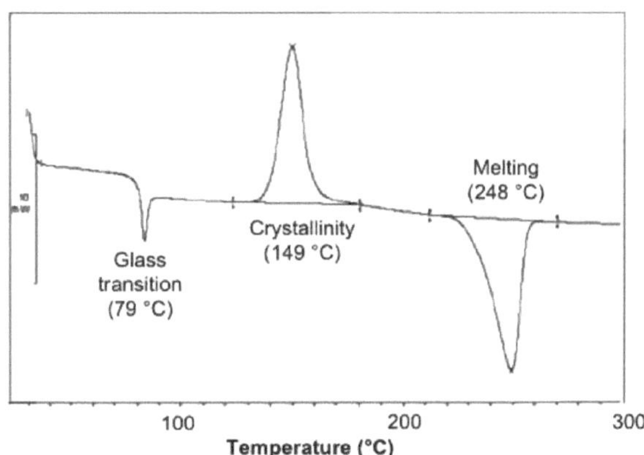

Figure 9.8: Standard exo up heat flux DSC spectrum of the PET polymer. Adapted from B. Demirel, A. Yaras, and H. Elçiçek, Crystallization behavior of PET materials. *BAÜ Fen Bil. Enst. Dergisi. Cilt*, 2011, 13, 26. Copyright: Balıkesir Üniversitesi (2011).

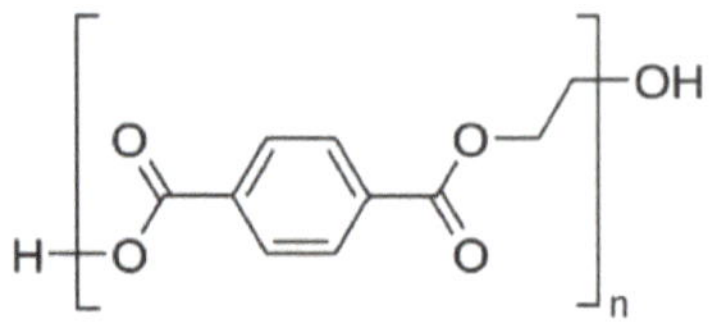

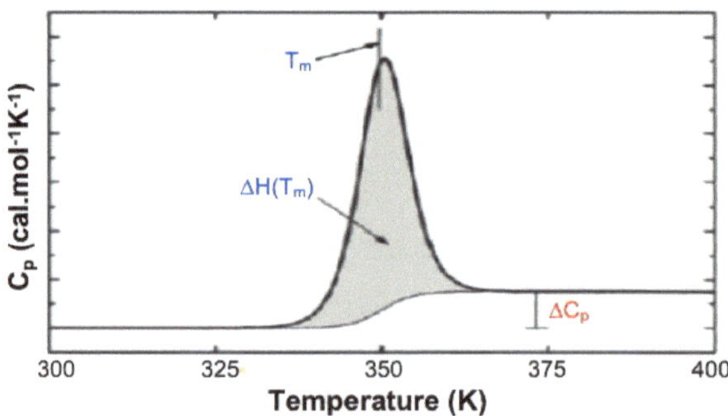

Figure 9.9: Structure of poly(ethylene terephthalate) (PET).

Heat capacity (C_p)

As previously stated, a typical plot created via DSC will be a measure of heat flow versus temperature. If the polymer undergoes no thermal processes, the plot of heat flow versus temperature will be zero slope. If this is the case, then the heat capacity of the polymer is proportional to the distance between the zero-slopped line and the x-axis. However, in most instances, the heat capacity is measured to be the slope of the resulting heat flow versus temperature plot. Note that any thermal alteration to a polymer will result in a change in the polymer's heat capacity; therefore, all DSC plots with a non-zero slope indicate some thermal event must have occurred.

However, it is also possible to directly measure the heat capacity of a polymer as it undergoes some phase change. To do so, a heat capacity versus temperature plot is to be created. In doing so it becomes easier to zero in on and analyze a weak thermal event in a reproducible manner. To measure heat capacity as a function of increasing temperature, it is necessary to divide all values of a standard DSC plot by the measured heating rate.

Figure 9.10: Direct DSC heat capacity measurement of a phase change material at the melting temperature. Reproduced from P. Giri and C. Pal, An overview on the thermodynamic techniques used in food chemistry. *Mod. Chem. Appl.*, 2014, 2, 142. Copyright: Longdom Publishing (2014).

For example, say a polymer has undergone a subtle thermal event at a relatively low temperature. To confirm a thermal event is occurring, zero in on the temperature range the event was measured to have occurred at and create a heat capacity versus temperature plot. The thermal event becomes immediately identifiable by the presence of a change in the polymer's heat capacity as shown in Figure 9.10.

Glass transition temperature (T_g)

As a polymer is continually heated within the DSC system, it may reach the glass transition: a temperature range under which a polymer can undergo a reversible transition between a brittle or viscous state. The temperature at which this reversible transition can occur is understood to be the glass transition temperature (T_g); however, the transition does not occur suddenly at one temperature but, instead, transitions slowly across a range of temperatures.

Once a polymer is heated to the glass transition temperature, it will enter a molten state. Upon cooling the polymer, it loses its elastic properties and instead becomes brittle, like glass, due to a decrease in chain mobility. Should the polymer continue to be heated above the glass transition temperature, it will become soft due to increased heat energy inducing different forms of transitional and segmental motion within the polymer, promoting chain mobility. This allows the polymer to be deformed or molded without breaking.

Upon reaching the glass transition range, the heat capacity of the polymer will change, typically become higher. In turn, this will produce a change in the DSC plot. The system will begin heating the sample pan at a different rate than the reference pan to accommodate this change in the polymer's heat capacity. Figure 9.8 is an example of the glass transition as measured by DSC. The glass transition has been highlighted, and the glass transition temperature is understood to be the mid-point of the transitional range.

While the DSC instrument will capture a glass transition, the glass transition temperature cannot, in actuality, be exactly defined with a standard DSC. The glass transition is a property that is completely dependent on the extent that the polymer is heated or cooled. As such, the glass transition is dependent on the applied heating or cooling rate of the DSC system. Therefore, the glass transition of the same polymer can have different values when measured on separate occasions. For example, if the applied cooling rate is lower during a second trial, then the measured glass transition temperature will also be lower.

However, in having a general knowledge of the glass transition temperature, it becomes possible to hypothesize the polymers chain length and structure. For example, the chain length of a polymer will affect the number of Van der Waal or entangling chain interactions that occur. These interactions will in turn determine just how resistant the polymer is to increasing heat. Therefore, the temperature at which T_g occurs is correlated to the magnitude of chain interactions. In turn, if the glass transition of a polymer is consistently shown to occur quickly at lower temperatures, it may be possible to infer that the polymer has flexible functional groups that promote chain mobility.

Crystallization (T_c)

Should a polymer sample continue to be heated beyond the glass transition temperature range, it becomes possible to observe crystallization of the polymer sample. Crystallization is the process by which polymer chains form ordered arrangements with one another.

Essentially, before the glass transition range, the polymer does not have enough energy from the applied heat to induce mobility within the polymer chains; however, as heat is continually added, the polymer chains begin to have greater and greater mobility. The chains eventually undergo transitional, rotational, and segmental motion as well as stretching, disentangling, and unfolding. Finally, a peak temperature is reached, and enough heat energy has been applied to the polymer that the chains are mobile enough to move into very ordered parallel, linear arrangements. At this point, crystallization begins. The temperature at which crystallization begins is the crystallization temperature (T_c).

As the polymer undergoes crystalline arrangements, it will release heat since intramolecular bonding is occurring. Because heat is being released, the process is exothermic, and the DSC system will lower the amount of heat being supplied to the sample plate in relation to the reference plate so as to maintain a constant temperature between the two plates. As a result, a positive amount of energy is released to the environment and an increase in heat flow is measured in an *exo* up DSC system, as seen in Figure 9.8.

The maximum point on the curve is known to be the T_c of the polymer while the area under the curve is the latent energy of crystallization, i.e., the change in the heat content of the system associated with the amount of heat energy released by the polymer as it undergoes crystallization. The degree to which crystallization can be measured by the DSC is dependent not only on the measured conditions but also on the polymer itself. For example, in the case of a polymer with very random ordering, i.e., an amorphous polymer, crystallization will not even occur.

In knowing the crystallization temperature of the polymer, it becomes possible to hypothesize on the polymer's chain structure, average molecular weight, tensile strength, impact strength, resistance to solvents, etc. For example, if the polymer tends to have a lower crystallization temperature and a small latent heat of crystallization, it becomes possible to assume that the polymer may already have a chain structure that is highly linear since not much energy is needed to induce linear crystalline arrangements. In turn, in obtaining crystallization data via DSC, it becomes possible to determine the percentage of crystalline structures within the polymer, or, the degree of crystallinity. To do so, compare the latent heat of crystallization, as determined by the area under the crystallization curve, to the latent heat of a standard sample of the same polymer with a known crystallization degree.

Knowledge of the polymer sample's degree of crystallinity also provides an avenue for hypothesizing the composition of the polymer. For example, having a very high degree of crystallinity may suggest that the polymer contains small, brittle molecules that are very ordered.

Melting behavior (T_m)

As the heat being applied pushes the temperature of the system beyond T_c, the polymer begins to approach a thermal transition associated with melting. In the melting phase, the heat applied provides enough energy to, now, break apart the intramolecular bonds holding together the crystalline structure, undoing the polymer chains' ordered arrangements. As this occurs, the temperature of the sample plate does not change as the applied heat is no longer being used to raise the temperature but instead to break apart the ordered arrangements.

As the sample melts, the temperature slowly increases as less and less of the applied heat is needed to break apart crystalline structures. Once all the polymer chains in the sample are able to move around freely, the temperature of the sample is said to reach the melting temperature (T_m). Upon reaching the melting temperature, the applied heat begins exclusively raising the temperature of the sample; however, the heat capacity of the polymer will have increased upon transitioning from the solid crystalline phase to the melt phase, meaning the temperature will increase more slowly than before.

Since, during the endothermic melting process of the polymer, most of the applied heat is being absorbed by the polymer, the DSC system must substantially increase the amount of heat applied to the sample plate so as to maintain the temperature between the sample plate and the reference plate. Once the melting temperature is reached, however, the applied heat of the sample plate decreases to match the applied heat of the reference plate. As such, since heat is being absorbed from the environment, the resulting *exo* up DSC plot will have a negative curve as seen in Figure 9.8 where the lowest point is understood to be the melt phase temperature. The area under the curve is, in turn, understood to be the latent heat of melting, or, more precisely, the change in the heat content of the system associated with the amount of heat energy absorbed by the polymer to undergo melting.

Once again, in knowing the melting range of the polymer, insight can be gained on the polymer's average molecular weight, composition, and other properties. For example, the greater the molecular weight or the stronger the intramolecular attraction between functional groups within crosslinked polymer chains, the more heat energy that will be needed to induce melting in the polymer.

Modulated DSC: an overview

While standard DSC is useful in characterization of polymers across a broad temperature range in a relatively quick manner and has user-friendly software, it still has a series of limitations with the main limitation being that it is highly operator dependent. These limitations can, at times, reduce the accuracy of analysis regarding the measurements of T_g, T_c and T_m, as described in the previous section. For example, when using a synthesized polymer that is composed of multiple blends of different monomer compounds, it can become difficult to interpret the various transitions of the polymer due to overlap. In turn, some transitional events are completely dependent on what the user decides to input for the heating or cooling rate.

To resolve some of the limitations associated with standard DSC, there exists modulated DSC (MDSC). MDSC not only uses a linear heating rate like standard DSC, but also uses a sinusoidal, or modulated, heating rate. In doing so, it is as though the MDSC is performing two, simultaneous experiments on the sample.

What is meant by a modulated heating rate is that the MDSC system will vary the heating rate of the sample by a small range of heat across some modulating period of time. However, while the temperature rate of change is sinusoidal, it is still ultimately increasing across time as indicated in Figure 9.11. In turn, Figure 9.12 also shows the sinusoidal heating rate as a function of time overlaying the linear heating rate of standard DSC. The linear heating rate of DSC is 2 °C/min and the modulated heating rate of MDSC varies from roughly ~0.1 °C/min and ~3.8 °C/min across a period of time.

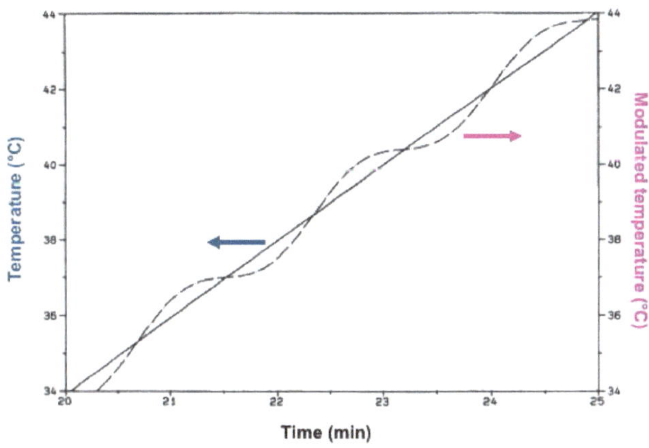

Figure 9.11: Schematic of sample temperature as a function of time with an underlying linear heating rate of Standard DSC. Adapted from E. Verdonck, K. Schaap, and L. C. Thomas, A discussion of the principles and applications of modulated temperature DSC (MTDSC). *Int. J. Pharm.*, **1999, 192, 3. Copyright: Elsevier (1999).**

By providing two heating rates, a linear and a modulated one, MDSC is able to measure more accurately how heating rates affect the rate of heat flow within a polymer sample. As such, MDSC offers a means to eliminate the applied heating rate aspects of operator dependency.

In turn, the MDSC instrument also performs mathematical processes that separate the standard DSC plot into reversing and non-reversing components. The reversing signal is representative of properties that respond to temperature modulation and heating rate, such as glass transition and melting. On the other hand, the non-reversing component is representative of kinetic, time-dependent process such as decomposition, crystallization, and curing. Figure 9.13 provides an example of such a plot using PET.

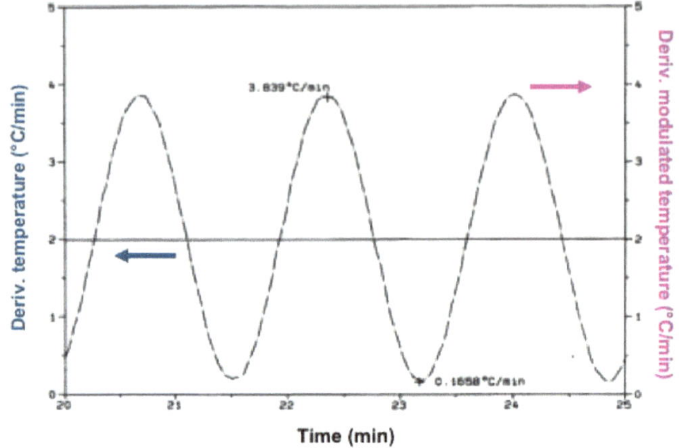

Figure 9.12: Modulated DSC signals of PET, split into reversing and non-reversing components as well as the total heat flow, showcasing the transitional change. Adapted from E. Verdonck, K. Schaap, and L. C. Thomas, A discussion of the principles and applications of modulated temperature DSC (MTDSC). *Int. J. Pharm.*, **1999, 192, 3. Copyright: Elsevier (1999).**

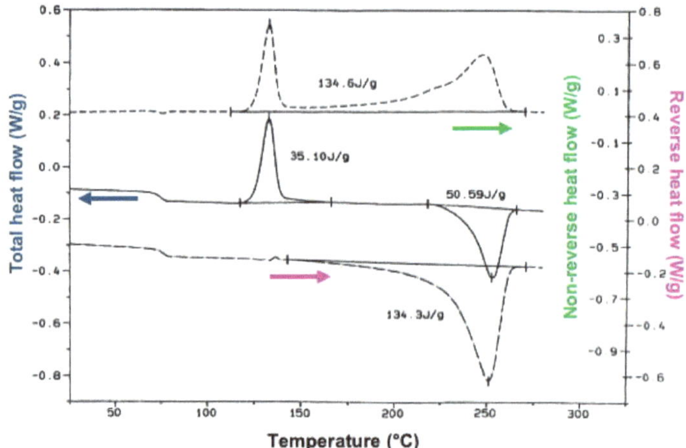

Figure 9.13: Modulated DSC signals of PET, split into reversing and non-reversing components as well as total heat flow, showcasing the related transitional temperatures. Adapted from E. Verdonck, K. Schaap, and L. C. Thomas, A discussion of the principles and applications of modulated temperature DSC (MTDSC). *Int. J. Pharm.*, **1999, 192, 3. Copyright: Elsevier (1999).**

The mathematics behind MDSC is most simply represented by:

$$dH/dt = C_p(dT/dt) + f(T,t)$$

where dH/dt is the total change in heat flow that would be derived from a standard DSC. C_p is heat capacity derived from modulated heating rate, dT/dt is representative of both the linear and modulated heating rate, and f(T,t) is representative of kinetic, time-dependent events, i.e., the non-reversing signal. When combining C_p and dT/dt, creating $C_p(dT/dt)$, the reversing signal is produced. The non-reversing signal is, therefore, found by simply subtracting the reversing signal from the total heat flow signal.

As such, MDSC is capable of independently measuring not only total heat flow but also the heating rate and kinetic components of said heat flow, meaning MDSC can break down complex or small transitions into their many singular components with improved sensitivity, allowing for more accurate analysis. Below are some cases in which MDSC proved to be useful for analytics.

Advanced analysis of T_g

Using a standard DSC, it can be difficult to ascertain the accuracy of measured transitions that are relatively weak, such as T_g, since these transitions can be overlapped by stronger, kinetic transitions. This is quite the problem as missing a weak transition could cause the misinterpretation of polymer to be a uniform sample as opposed to a polymer blend. To resolve this, it is useful to split the plot into its reversing component, i.e., the portion which will contain heat dependent properties like T_g, and its non-reversing, kinetic component.

For example, shown in the Figure 9.14 is the MDSC of an unknown polymer blend which, upon analysis, is composed of PET, amorphous polycarbonate (PC, Figure 9.15), and a high-density polyethylene (HDPE, Figures 2.11 and 9.16). Looking at the reversing signal, the T_g of polycarbonate is around 140 °C and the T_g of PET is around 75 °C. As seen in the total heat flow signal, which is representative of a standard DSC plot, the T_g of PC would have been more difficult to analyze and, as such, may have been incorrectly analyzed.

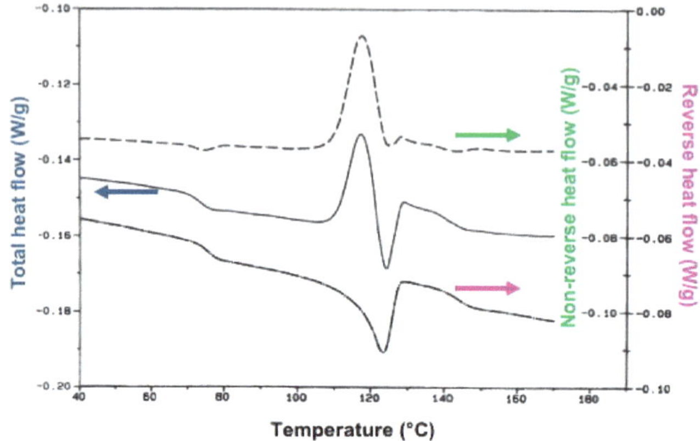

Figure 9.14: MDSC signals of a polymer blend composed of HDPE, PC, and PET. Adapted from E. Verdonck, K. Schaap, and L. C. Thomas, A discussion of the principles and applications of modulated temperature DSC (MTDSC). *Int. J. Pharm.*, **1999, 192, 3.**

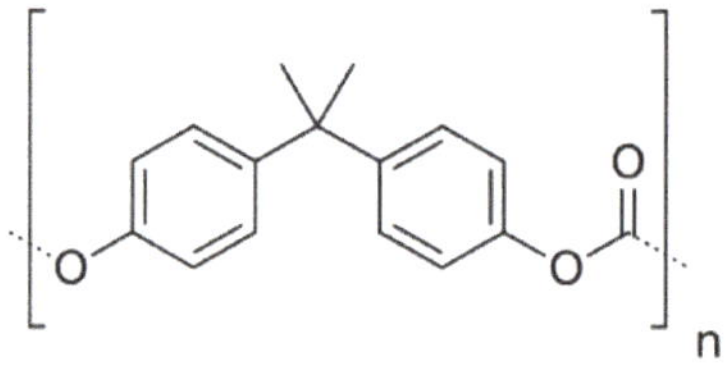

Figure 9.15: Structure of polycarbonate.

Figure 9.16: Beads of high-density polyethylene (HDPE).

Advanced analysis of T_m

Further, there are instances in which a polymer or, more likely, a polymer blend will produce two different sets of crystalline structures. With two crystalline structures, the resulting melting peak will be poorly defined and, thus, difficult to analyze via a standard DSC.

Using MDSC, however, it becomes possible to isolate the reversing signal, which will contain the melting curve. Through isolation of the reversing signal, it becomes clear that there is an overlapping of two melting peaks such that the MDSC system reveals two melting points. For example, as seen in Figure 9.16, the analysis of a poly(lactic acid) polymer (PLA, Figure 9.17) with 10 wt% of a plasticize (P600, Figure 9.18) reveals two melting peaks in the reversing signal not visible in the total heat flow. The presence of two melting peaks could, in turn, suggest the formation of two crystalline structures within the polymer sample. Other interpretations are, of course, possible via analyzing the reversing signal.

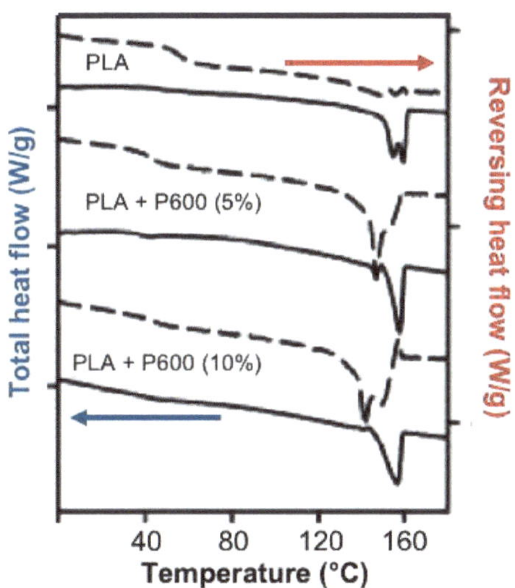

Figure 9.16: MDSC of poly(lactic acid) (PLA) with varying concentrations of a plasticizer. The solid lines represent total heat flow and the dashed lines represent reversing heat flow. Adapted from Z. Kulinski and E. Piorkowska, Crystallization, structure and properties of plasticized poly(l-lactide). *Polymer,* **2005, 46, 10290. Copyright: Elsevier (2005).**

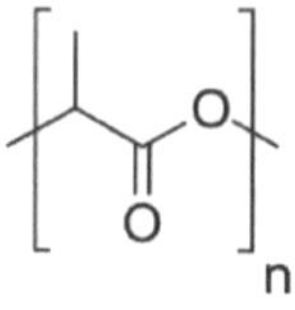

Figure 9.17: Structure of poly(lactic acid) (PLA).

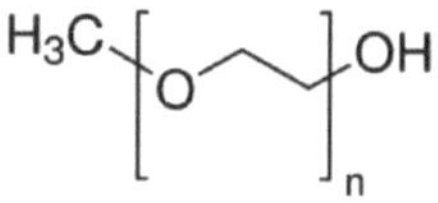

Figure 9.18: Structure of polyethylene glycol methyl ether.

Analysis of polymer aging

In many instances, polymers may be left to sit in refrigeration or stored at temperatures below their respective glass transition temperatures. By leaving a polymer under such conditions, the polymer is situated to undergo physical aging. Typically, the more flexible the chains of a polymer are, the more likely they will undergo time-related changes in storage. That is to say, the polymer will begin to undergo molecular relaxation such that the chains will form very dense regions while they conglomerate together. As the polymer ages, it will tend towards embrittlement and develop internal stresses. As such, it is very important to be aware if the polymer being studied has gone through aging while in storage.

If a polymer has undergone physical aging, it will develop a new endothermic peak when undergoing thermal analysis. This occurs because, as the polymer is being heated, the polymer chains absorb heat, increase mobility, and move to a more relaxed condition as time goes on, transforming back to pre-aged conditions. In turn an endothermic shift, in association with this heat absorbance, will occur just before the T_g step change. This peak is known as the enthalpy of relaxation (ΔH_R).

Since the T_g and ΔH_R are relatively close to one another energy-wise, they will tend to overlap, making it difficult to distinguish the two from one another. However, ΔH_R is a kinetics dependent thermal shift while T_g is a heating dependent thermal shift; therefore, the two can be separated into a non-reversing and reversing plot via MDSC and be independently analyzed.

Figure 9.14 is an example of an MDSC plot of a polymer blend of PET, PC, and HDPE in which the enthalpy of relaxation of PET is visible in the dashed non reversing signal around 75 °C. In turn, within the reversing signal, the glass transition of PET is visible around 75 °C as well.

Quasi-isothermal DSC

While MDSC is a strong step in the direction of eliminating operator error, it is possible to have an even higher level of precision and accuracy when analyzing a polymer. To do so, the DSC system must expose the sample to quasi-isothermal conditions. In creating quasi-isothermal conditions, the polymer sample is held at a specific temperature for extended periods of time with no applied heating rate. With the heating rate being effectively zero, the conditions are isothermal. The temperature of the sample may change, but the change will be derived solely from a kinetic transition that has occurred within the polymer. Once a kinetic transition has occurred within the polymer, it will absorb or release some heat, which will raise or decrease the temperature of the system without the application of any external heat.

In creating these conditions, issues created by the variation of the applied heating rate by operators is no longer a large concern. Further, in subjecting a polymer sample to quasi-isothermal conditions, it becomes possible to get improved and more accurate measurements of heat dependent thermal events, such as events typically found in the reversing signal, as a function of time.

Improved glass transition

As mentioned earlier, the glass transition is volatile in the sense that it is highly dependent on the heating and cooling rate of the DSC system as applied by the operator. A minor change in the heating or cooling rate between two experimental measurements of the same polymer sample can result in fairly different measured glass transitions, even though the sample itself has not been altered. Remember also, that the glass transition is a measure of the changing C_p of the polymer sample as it crosses certain heat energy thresholds. Therefore, it should be possible to capture a more accurate and precise glass transition under quasi-isothermal conditions since these conditions produce highly accurate C_p measurements as a function of time.

By applying quasi-isothermal conditions, the polymer's C_p can be measured in fixed-temperature steps within the apparent glass transition range as measured via standard DSC. In measuring the polymer across a set of quasi-isothermal steps, it becomes possible to obtain changing C_p rates that, in turn, would be nearly reflective of an exact glass transition range for a polymer.

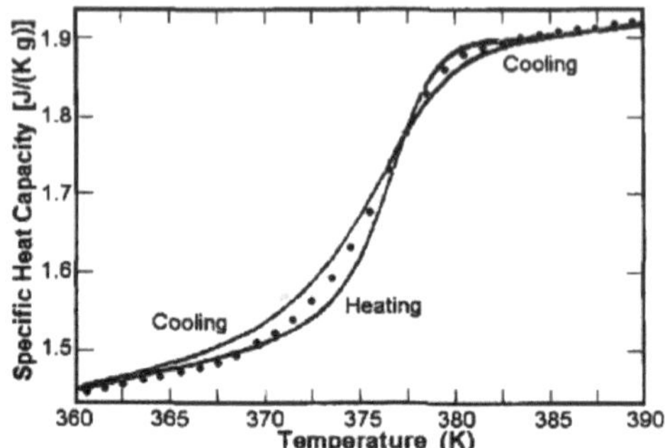

Figure 9.19: MDSC plot of polystyrene measuring the glass transition as changes in heat capacity as a function of temperature using either a heating or cooling rate as indicated by the solid lines. The dotted line indicates a quasi-isothermal measurement of the glass transition of polystyrene. Adapted from L. C. Thomas, A. Boller, I. Okazaki, and B. Wunderlich, Modulated differential scanning calorimetry in the glass transition region, IV. Pseudo-isothermal analysis of the polystyrene glass transition. *Thermochim. Acta*, 1997, 291, 85. Copyright: Elsevier (1997).

In Figure 9.19 the glass transition of polystyrene (Figure 2.1) is shown to vary depending on the heating or cooling rate of the DSC; however, when applying quasi-isothermal conditions and measuring the heat capacity at temperature steps produces a very accurate glass transition that can be used as a standard for comparison.

Bibliography

V. Berstein and V. Egorov, *Differential Scanning Calorimetry of Polymers: Physics, Chemistry, Analysis, Technology*, Ellis Horwood, New York, (1994).

Characterization of Polymers using Differential Scanning Calorimetry (DSC), White Paper EAG Laboratories. 2016

J. Dean, *The Analytical Chemistry Handbook*, McGraw Hill: New York, (1995).

B. Demirel, A. Yaras, and H. Elçiçek, Crystallization behavior of PET materials. *BAÜ Fen Bil. Enst. Dergisi. Cilt*, 2011, **13**, 26.

P. Giri and C. Pal, An overview on the thermodynamic techniques used in food chemistry. *Mod. Chem. Appl.*, 2014, 2, 142.

E. Gmelin and St. M. Sarge, Calibration of differential scanning calorimeters. *Pure Appl. Chem.*, 1995, **67**, 1789.

Heat Capacity Measurements Using Quasi-Isothermal MDSC, Thermal Analysis and Rheology. TA Instruments.

G. Hohne, W. Hemminger, and H. Flammersheim, *Differential Scanning Calorimetry*, 2nd edn., Springer, New York, (2003).

Z. Kulinski and E. Piorkowska, Crystallization, structure and properties of plasticized poly(l-lactide). *Polymer*, 2005, **46**, 10290.

L. C. Thomas, A. Boller, I. Okazaki, and B. Wunderlich, Modulated differential scanning calorimetry in the glass transition region, IV. Pseudo-isothermal analysis of the polystyrene glass transition. *Thermochim. Acta*, 1997, **291**, 85.

E. Verdonck, K. Schaap, and L. C. Thomas, A discussion of the principles and applications of modulated temperature DSC (MTDSC). *Int. J. Pharm.*, 1999, **192**, 3.

E. Watson, M. O'Neill, J. Justin, and N. Brenner, A differential scanning calorimeter for quantitative differential thermal analysis. *Anal. Chem.*, 1964, **36**, 1233.

E. Watson and M. O'Neill, US Patent 3,263,484 (1966).

Chapter 10: Thermal Conductivity

Juan Velazquez and Andrew R. Barron

Introduction

Principles of thermal conductivity

Thermal conductivity is the ability of a chemical specie to conduct heat. Each gas has a different thermal conductivity. The units of thermal conductivity in the international system of units are W/m·K. Table 10.1, shows the thermal conductivity of some common gasses.

Gas	Thermal conductivity (W/m·K)
Hydrogen	0.18050
Argon	0.01772
Helium	0.15130
Carbon monoxide	0.02614

Table 10.1: Thermal conductivity values for common gasses.

Catalysis characterization: a case study

A catalyst is a "substance that accelerates the rate of chemical reactions without being consumed". Some reactions, such as the hydrodechlorination of TCE,

$$C_2Cl_3H + 4H_2 \xrightarrow{Pd} C_2H_6 + 3HCl$$

does not occur spontaneously but can occur in the presence of a catalyst.

Metal dispersion is a common term within the catalyst industry. The term refers to the amount of metal that is active for a specific reaction. Let's assume a catalyst material has a composition of 1 wt% palladium and 99% alumina (Al_2O_3) (Figure 10.2). Even though the catalyst material has 1 wt% of palladium, not all the palladium is active. The material might be oxidized due to air exposure or some of the material is not exposed to the surface (Figure10.3), hence it can't participate in the reaction. For this reason, it is important to characterize the material.

Figure 10.2: A photograph of a sample of commercially available 1 wt% Pd/Al₂O₃.

Figure 10.3: Representation of Pd nanoparticles on Al₂O₃. Some palladium atoms are exposed to the surface, while some other lay below the surface atoms and are not accessible for reaction.

In order for Pd to react it needs to be in the metallic form. Any oxidized palladium will be inactive. Thus, it is important to determine the oxidation state of the Pd atoms on the surface of the material. This can be accomplished using an experiment called temperature programmed reduction (TPR). Subsequently, the percentage of active palladium can be determined by hydrogen chemisorption. The percentage of active metal is an important parameter when comparing the performance of multiple catalyst. Usually the rate of reaction is normalized by the amount of active catalyst.

Thermal conductivity detection

A thermal conductivity detector has four laments that change resistance according to the thermal conductivity of the gas flowing over it. Two laments measure the reference gas and the other two measures the sample gas. The detector is isothermal; it will increase or decrease the voltage in each of the resistors in order to maintain a constant temperature. The temperature of the detector is 125 °C. When both the reference and samples gas have the same composition and same flow rate, the resistors are balanced, and the detector will zero the signal. If there is a change in flow rate or in the gas composition the detector will react to maintain the constant temperature. The detector circuitry can be described using a Wheatstone bridge configuration as shown in Figure 10.4. If the gas flowing through the sample has a higher thermal conductivity the lament will cool down, the detector will apply a higher voltage to keep a constant temperature and this will be recorded as a positive signal.

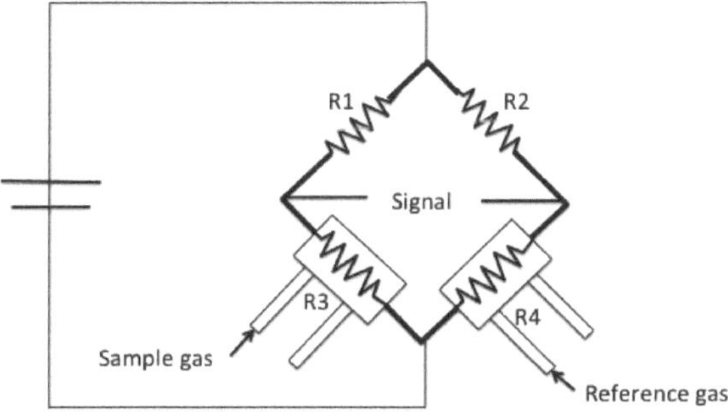

Figure 10.4: A simplified circuit diagram of a thermal conductivity detector.

This detector is part of a typical commercial instrument such as a Micromeritics AutoChem 2920 (Figure 10.5). This instrument is an automated analyzer with the ability to perform chemical adsorption and temperature-programmed reactions on a catalyst, catalyst support, or other materials.

Figure 10.5: A photograph of a Micromeritics AutoChem 2920.

Temperature programmed reduction (TPR)

TPR will determine the number of reducible species on a catalyst and will tell at what temperature each of these species was reduced. For example palladium is ordinarily found as Pd(0) or Pd(II), i.e., oxidation states 0 and +2. Pd(II) can be reduced at very low temperatures (5 – 10 °C) to Pd(0) following:

$$PdO + H_2 \rightarrow Pd(0) + H_2O$$

A 128.9 mg 1wt% Pd/Al$_2$O$_3$ samples is used for the experiment, Figure 10.6. Since we want to study the oxidation state of the commercial catalyst, no pre-treatment needs to be executed to the sample. A 10% hydrogen-argon mixture is used as analysis and reference gas. Argon has a low thermal conductivity and hydrogen has a much higher thermal conductivity. All gases will flow at 50 cm^3/min. The TPR experiment will start at an initial temperature of 200 K, temperature ramp 10 K/min, and final temperature of 400 K. The H$_2$/Ar mixture is flowed through the sample, and past the detector in the analysis port. While in the reference port the mixture doesn't become in contact with the sample. When the analysis gas starts flowing over the sample, a baseline reading is established by the detector. The baseline is established at the initial temperature to ensure there is no reduction. While this gas is flowing, the temperature of the sample is increased linearly with time and the consumption of hydrogen is recorded. Hydrogen atoms react with oxygen atoms to form H$_2$O.

Figure 10.6: A sample of Pd/Al$_2$O$_3$ in a typical sample holder.

Water is removed from the gas stream using a cold trap. As a result, the amount of hydrogen in the argon/hydrogen gas mixture decreases and the thermal conductivity of the mixture also decrease. The change is compared to the reference gas and yields to a hydrogen uptake volume. Figure 10.7 is a typical TPR profile for PdO.

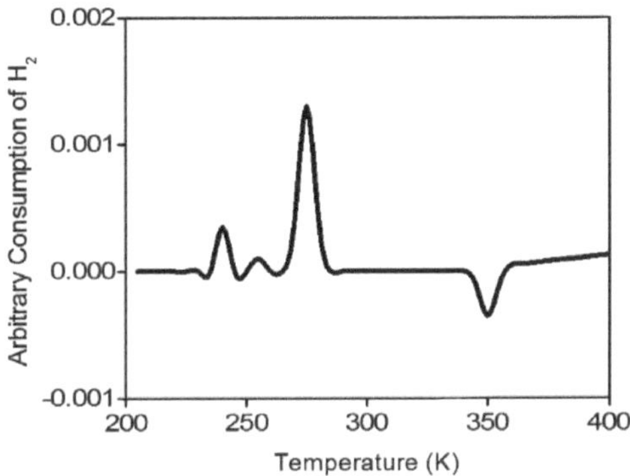

Figure 10.7: A typical TPR profile of PdO. Adapted from R. Zhang, J. A. Schwarz, A. Datye, and J. P. Baltrus, The effect of second-phase oxides on the catalytic properties of dispersed metals: palladium supported on 12% WO₃/Al₂O₃. *J. Catal.*, 1992, 138, 55. Copyright: Elsevier (1992).

Pulse chemisorption

Once the catalyst (1 wt% Pd/Al₂O₃) has been completely reduced, the user will be able to determine how much palladium is active. A pulse chemisorption experiment will determine active surface area, percentage of metal dispersion and particle size. Pulses of hydrogen will be introduced to the sample tube in order to interact with the sample. In each pulse hydrogen will undergo a dissociative adsorption on to palladium active sites until all palladium atoms have reacted. After all active sites have reacted, the hydrogen pulses emerge unchanged from the sample tube. The amount of hydrogen chemisorbed is calculated as the total amount of hydrogen injected minus the total amount eluted from the system.

Data collection for hydrogen pulse chemisorption

The sample from previous experiment (TPR) will be used for this experiment. Ultra-high purity argon will be used to purge the sample at a flow rate of 40 cm³/min. The sample will be heated to 200 °C in order to remove all chemisorbed hydrogen atoms from the Pd(0) surface. The sample is cooled down to 40 °C. Argon will be used as carrier gas at a flow of 40 cm^3/min. Filaments temperature will be 175 °C and the detector temperature will be 110 °C. The injection loop has a volume of 0.03610 cm^3 @ STP. As shown in Figure 10.8., hydrogen pulses will be injected into the flow stream, carried by argon to become in contact and react with the sample. It should be noted that the first pulse of hydrogen was almost completely adsorbed by the sample. The second and third pulses show how the samples is been saturated. The positive value of the TCD detector is consistent with our assumptions. Since hydrogen has a higher thermal conductivity than argon, as it flows through the detector it will tend to cool down the laments, the detector will then apply a positive voltage to the laments in order to maintain a constant temperature.

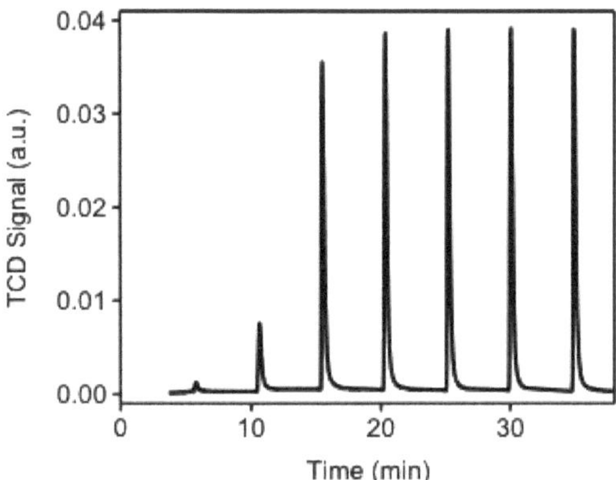

Figure 10.8: A typical hydrogen pulse chemisorption profile of 1 wt% Pd/Al₂O₃.

Pulse chemisorption data analysis

Table 10.2 shows the integration of the peaks from Figure 10.8. This integration is performed by an automated software provided with the instrument. It should be noted that the first pulse is completely consumed by the sample, the pulse was injected between time 0 and 5 minutes. From Figure 10.8 it is observed that during the first four pulses, hydrogen is consumed by the sample. After the fourth pulse, it appears the sample is not consuming hydrogen. The experiment continues for a total of seven pulses, at this point the software determines that no consumption is occurring and stops the experiment. Pulse eight is denominated the "saturation peak", meaning the pulse at which no hydrogen was consumed.

Pulse n	Area
1	0
2	0.000471772
3	0.00247767
4	0.009846683
5	0.010348201
6	0.010030243
7	0.009967717
8	0.010580979

Table 19.2: Hydrogen pulse chemisorption data.

The change in area ($\Delta area_n$) is calculated for each peak pulse area ($area_n$) is calculated using,

$$\Delta area_n = area_{saturation} - area_n$$

and compared to that of the saturation pulse area ($area_{saturation}$ = 0.010580979). Each of these changes in area is proportional to an amount of hydrogen consumed by the sample in each pulse. Table 10.3 Shows the calculated change in area.

Pulse n	$area_n$	$\Delta area_n$
1	0	0.010580979
2	0.000471772	0.0105338018
3	0.00247767	0.008103309
4	0.009846683	0.000734296
5	0.010348201	0.000232778
6	0.010030243	0.000550736
7	0.009967717	0.000613262
8	0.010580979	0

Table 10.3: Hydrogen pulse chemisorption data with Δarea.

The $\Delta area_n$ values are then converted into hydrogen gas consumption using,

$$V_{adsorbed} = \frac{\Delta area_n \times F_c}{SW}$$

where F_c is the area- to-volume conversion factor for hydrogen and SW is the weight of the sample. F_c is equal to 2.6465 cm^3/peak area. Table 10.4 shows the results of the volume adsorbed and the cumulative volume adsorbed. Using the data on Table 10.4, a series of calculations can now be performed in order to have a better understanding of our catalyst properties.

Pulse n	area$_n$	Δarea$_n$	V adsorbed (cm^3/g STP)	Cumulative quantity (cm^3/g STP)
1	0	0.0105809790	0.2800256	0.2800256
2	0.000471772	0.0105338018	0.2787771	0.5588027
3	0.00247767	0.0081033090	0.2144541	0.7732567
4	0.009846683	0.0007342960	0.0194331	0.7926899
5	0.010348201	0.0002327780	0.0061605	0.7988504
6	0.010030243	0.0005507360	0.0145752	0.8134256
7	0.009967717	0.0006132620	0.0162300	0.8296556
8	0.010580979	0	0.0000000	0.8296556

Table 10.4: Includes the volume adsorbed per pulse and the cumulative volume adsorbed.

Gram molecular weight

Gram molecular weight is the weighted average of the number of moles of each active metal in the catalyst. Since this is a monometallic catalyst, the gram molecular weight is equal to the molecular weight of palladium (106.42 g/mol). The GMC$_{Calc}$ is calculated using,

$$GMW_{Calc} = \frac{1}{\left(\frac{F_1}{W_{atomic1}}\right) + \left(\frac{F_2}{W_{atomic2}}\right) + ... + \left(\frac{F_N}{W_{atomicN}}\right)}$$

where F is the fraction of sample weight for metal N and $W_{atomicN}$ is the gram molecular weight of metal N (g/g-mole). The calculation for this experiment is shown in,

$$GMW_{Calc} = \frac{1}{\left(\frac{F_1}{W_{atomicPd}}\right)} = \frac{W_{atomicPd}}{F_1} = \frac{106.42\frac{g}{g\text{-mole}}}{1} = 106.42\frac{g}{g\text{-mole}}$$

Metal dispersion

The metal dispersion is calculated using,

$$PD = 100 \times \left(\frac{V_s \times SF_{Calc}}{SW \times 22414} \right) \times GMW_{Calc}$$

where PD is the percent metal dispersion, V_s is the volume adsorbed (cm^3 at STP), SF$_{Calc}$ is the calculated stoichiometry factor (equal to 2 for a palladium-hydrogen system), SW is the sample weight and GMW$_{calc}$ is the calculated gram molecular weight of the sample (g/g-mol). Therefore, we obtain a metal dispersion of 6.03%:

$$PD = 100 \times \left(\frac{0.8296556 \, [cm^3] \times 2}{0.1289 \, [g] \times 22414 \left[\frac{cm^3}{mol} \right]} \right) \times 106.42 \left[\frac{g}{g\text{-}mol} \right] = 6.03\%$$

Metallic surface area per gram of metal

The metallic surface area per gram of metal is calculated using,

$$SA_{Metallic} = \left(\frac{V_s}{SW_{Metal} \times 22414} \right) \times (SF_{Calc}) \times (6.022 \times 10^{23}) \times SA_{Pd}$$

where SA$_{metallic}$ is the metallic surface area (m^2/g of metal), SW$_{Metal}$ is the active metal weight, SF$_{calc}$ is the calculated stoichiometric factor and SA$_{Pd}$ is the cross-sectional area of one palladium atom (nm^2). Thus, we obtain a metallic surface area of 2420.99 m^2/g-metal from:

$$SA_{Metallic} = \left(\frac{0.8296556 \, [cm^3]}{0.001289 \, [g_{metal}] \times 22414 \left[\frac{cm^3}{mol} \right]} \right) \times (2) \times \left(6.022 \times 10^{23} \left[\frac{atoms}{mol} \right] \right) \times 0.07 \left[\frac{nm^2}{atom} \right] = 2420.99 \left[\frac{m^2}{g\text{-}metal} \right]$$

Active particle size

The active particle size is estimated using,

$$APS = \frac{6}{D_{Calc} \times \left(\frac{W_s}{GMW_{Calc}} \right) \times (6.022 \times 10^{23}) \times SA_{Metallic}}$$

where D_{calc} is palladium metal density (g/cm^3), SW_{metal} is the active metal weight, GMW_{calc} is the calculated gram molecular weight (g/g-mole), and SA_{Pd} is the cross-sectional area of one palladium atom (nm^2). Thus, it is possible to obtain an optical particle size of 2.88 nm:

$$APS = \frac{600}{\left(1.202{\times}10^{-20}\left[\frac{g_{Pd}}{nm^3}\right]\right) \times \left(\frac{0.001289[g]}{106.42\left[\frac{g_{Pd}}{mol}\right]}\right) \times \left(6.022{\times}10^{23}\left[\frac{atoms}{mol}\right]\right) \times \left(2420.99\left[\frac{m^2}{g_{Pd}}\right]\right)} = 2.88\,nm$$

In a commercial instrument, a summary report will be provided which summarizes the properties of our catalytic material, e.g., Table 10.5.

Properties	Value
Palladium atomic weight	106.4 g/mol
Atomic cross-sectional area	0.0787 nm^2
Metal density	12.02 g/cm^3
Palladium loading	1 wt%
Metal dispersion	6.03%
Metallic surface area	2420.99 m^2/g-metal
Active particle diameter (hemisphere)	2.88 nm

Table 10.5: An example of the data in a summary report provided by Micromeritics AuthoChem 2920.

Bibliography

A. J. Canty, Development of organopalladium(IV) chemistry: fundamental aspects and systems for studies of mechanism in organometallic chemistry and catalysis. *Acc. Chem. Res.*, 1992, **25**, 83.

H. S. Fogler, *Elements of Chemical Reaction Engineering*, Prentice-Hall, New York (1992).

AutoChem 2920 Automated catalyst characterization system Operators Manual, Micromeritics Instrument Corporation, V4.01 (2009).

M. O. Nutt, K. N. Heck, P. Alvarez, and M. S. Wong, Improved Pd-on-Au bimetallic nanoparticle catalysts for aqueous-phase trichloroethene hydrodechlorination. *Appl. Catal. B-Environ.*, 2006, **69**, 115.

M. O. Nutt, J. B. Hughes, and M. S. Wong, Designing Pd-on-Au bimetallic nanoparticle catalysts for trichloroethene hydrodechlorination. *Environ. Sci. Technol.*, 2005, **39**, 1346.

P. A. Webb and C. Orr, *Analytical Methods in Fine Particle Technology*, Micromeritics Instrument Corp, 1997.

R. Zhang, J. A. Schwarz, A. Datye, and J. P. Baltrus, The effect of second-phase oxides on the catalytic properties of dispersed metals: palladium supported on 12% WO_3/Al_2O_3. *J. Catal.*, 1992, **138**, 55.

Chapter 11: Temperature-Programmed Desorption

Chih-Chau Hwang and Andrew R. Barron

Introduction

The temperature-programmed desorption (TPD) technique is often used to monitor surface interactions between adsorbed molecules and substrate surface. Utilizing the dependence on temperature is able to dis- criminate between processes with different activation parameters, such as activation energy, rate constant, reaction order and Arrhenius pre-exponential factor in order to provide an example of the set-up and results from a TPD experiment we are going to use an ultra-high vacuum (UHV) chamber equipped with a quadrupole mass spectrometer to exemplify a typical surface gas-solid interaction and estimate several important kinetic parameters.

Experimental system

Ultra-high vacuum (UHV) chamber

When we start to set up an apparatus for a typical surface TPD experiment, we should first think about how we can generate an extremely clean environment for the solid substrate and gas adsorbents. Ultra-high vacuum (UHV) is the most basic requirement for surface chemistry experiments. UHV is defined as a vacuum regime lower than 10^{-9} Torr. At such a low pressure the mean free path of a gas molecule is approximately 40 Km, which means gas molecules will collide and react with sample substrate in the UHV chamber many times before colliding with each other, ensuring all interactions take place on the substrate surface.

Most of time UHV chambers require the use of unusual materials in construction and by heating the entire system to ~180 °C for several hours baking to remove moisture and other trace adsorbed gases around the wall of the chamber in order to reach the ultra-high vacuum environment. Also, outgas from the substrate surface and other bulk materials should be minimized by careful selection of materials with low vapor pressures, such as stainless steel, for everything inside the UHV chamber. Thus, bulk metal crystals are chosen as substrates to study interactions between gas adsorbates and crystal surface itself. Figure 11.1 shows a schematic of a TPD system, while Figure 11.2

shows a typical TPD instrument equipped with a quadrupole MS spectrometer and a reflection absorption infrared spectrometer (RAIRS).

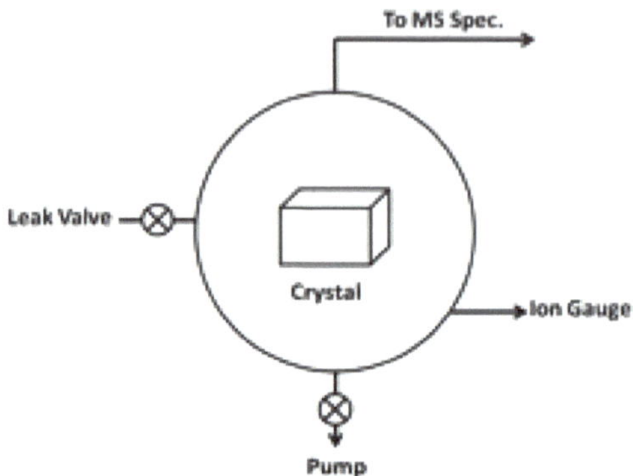

Figure 11.1: Schematic diagram of a TPD apparatus.

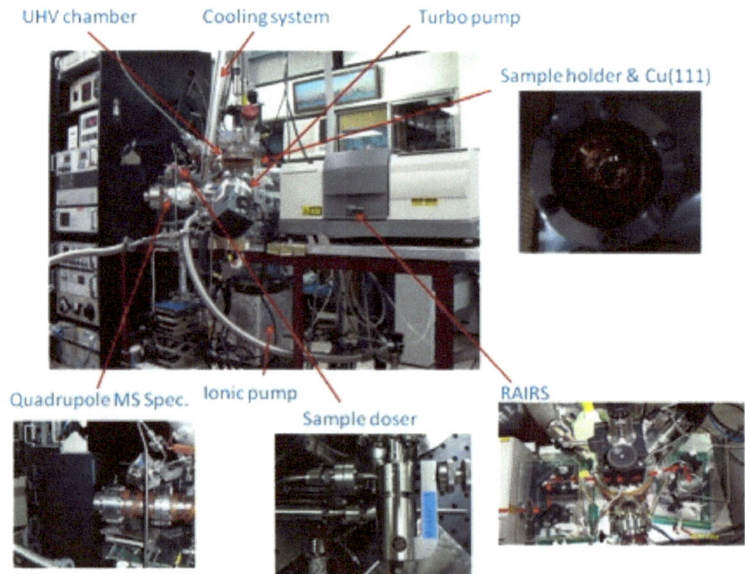

Figure 11.2: A typical TPD apparatus composed of a UHV chamber equipped with a serious of pumping systems, cooling system, sample dosing system as well as surface detection instruments including a quadrupole mass spectrometer and a reflection absorption infra-red spectrometer (RAIRS).

Pumping system

There is no single pump that can operate all the way from atmospheric pressure to UHV. Instead, a series of different pumps are used, according to the appropriate pressure range for each pump. Pumps are commonly used to achieve UHV include:

- Turbomolecular pumps (turbo pumps).
- Ionic pumps.
- Titanium sublimation pumps.
- Non-evaporate mechanical pumps.

UHV pressures are measured with an ion gauge, either a hot lament or an inverted magnetron type. Finally, special seals and gaskets must be used between components in a UHV system to prevent even trace leakage. Nearly all such seals are all metal, with knife edges on both sides cutting into a soft (e.g., copper) gasket. This all-metal seal can maintain system pressures down to $\sim 10^{-12}$ Torr.

Manipulator and bulk metal crystal

A UHV manipulator (or sample holder, see Figure 11.2) allows an object that is inside a vacuum chamber and under vacuum to be mechanically positioned. It may provide rotary motion, linear motion, or a combination of both. The manipulator may include features allowing additional control and testing of a sample, such as the ability to apply heat, cooling, voltage, or a magnetic field. Sample heating can be accomplished by thermal radiation. A lament is mounted close to the sample and resistively heated to high temperature. In order to simplify complexity from the interaction between substrate and adsorbates, surface chemistry labs often carry out TPD experiments by choosing a substrate with single crystal surface instead of polycrystalline or amorphous substrates (see Figure 11.2).

Pretreatment

Before the selected gas molecules are dosed to the chamber for adsorption, the substrates (metal crystals) need to be cleaned through argon plasma sputtering, followed by annealing at high temperature for surface reconstruction. After these pretreatments, the system is again cooled down to very low temperature (liquid N_2 temp), which facilitating gas molecules adsorbed on the substrate surface. Adsorption is a process in which a molecule becomes adsorbed onto a surface of another phase. It is distinguished from absorption, which is used when describing uptake into the bulk of a solid or liquid phase.

Temperature-programmed desorption processes

After gas molecules adsorption, now we are going to release theses adsorbates back into gas phase by programmed-heating the sample holder. A mass spectrometer is set up for collecting these desorbed gas molecules, and then correlation between desorption temperature and fragmentation of desorbed gas molecules will show us certain important information. Figure 11.3 shows a typical TPD experiment carried out by adsorbing CO onto Pd(111) surface, followed by programmed-heating to desorb the CO adsorbates.

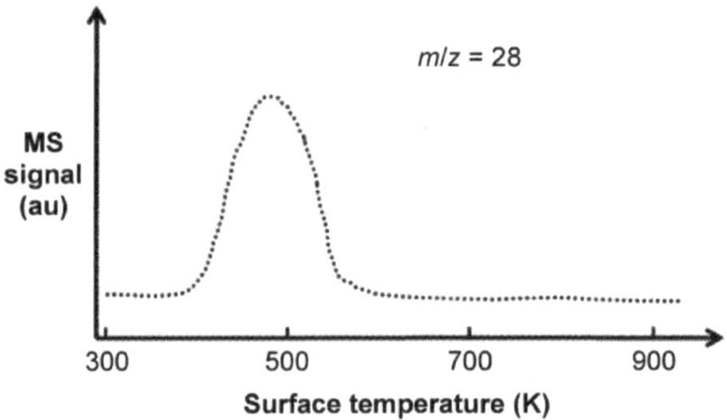

Figure 11.3: MS spectrum taken from a TPD experiment that CO ($m/z = 28$) was first adsorbed on Pd(111) surface, followed by desorbing at a fixed heating rate. The desorption rate which is proportional to the MS signal reaches its maximum around 500 K.

Theory of the TPD experiment

Langmuir isotherm

The Langmuir isotherm describes the dependence of the surface coverage of an adsorbed gas on the pressure of the gas above the surface at a fixed temperature. Langmuir isotherm is the simplest assumption, but it provides a useful insight into the pressure dependence of the extent of surface adsorption. It was Irving Langmuir who first studied the adsorption process quantitatively. In his proposed model, he supposed that molecules can adsorb only at specific sites on the surface, and that once a site is occupied by one molecule, it cannot adsorb a second molecule. The adsorption process can be represented as,

$$A + S \rightarrow A\text{-}S$$

where A is the adsorbing molecule, S is the surface site, and A-S stands for an A molecule bound to the surface site. In a similar way, reverse desorption process can be represented as:

$$A\text{-}S \rightarrow A + S$$

According to the Langmuir model, we know that the adsorption rate should be,

$$\text{adsorption} = k_a[A](1\text{-}\theta)$$

where θ is the fraction of the surface sites covered by adsorbate A. The desorption rate is then,

$$\text{desorption} = k_d\theta$$

K_a and k_d are the rate constants for the adsorption and desorption. At equilibrium, the rates of these two processes are equal, i.e.,

$$k_a[A](1\text{-}\theta) = k_d\theta$$

$$\frac{\theta}{1\text{-}\theta} = \frac{k_a}{k_d}[A]$$

$$K = k_a/k_d$$

$$\theta = \frac{K[A]}{1+K[A]}$$

We can replace [A] by P, where P means the gas partial pressure,

$$\theta = \frac{KP}{1+KP}$$

We can observe the equation above and know that if [A] or P is low enough so that K[A] or KP $\ll$ 1, then $\theta \sim$ K[A] or KP, which means that the surface coverage should increase linearly with [A] or P. On the contrary, if[A] or P is large enough so that K[A] or KP $\gg$ 1, then $\theta\sim$1. This behavior is shown in the plot of θ versus [A] or P, in Figure 11.4.

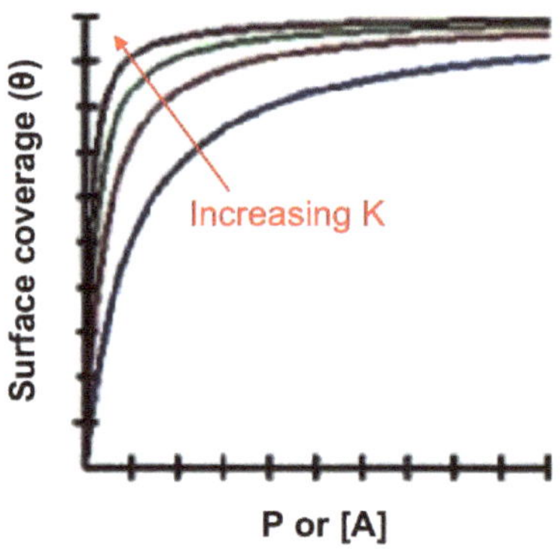

Figure 11.4: Simulated Langmuir isotherms. Value of constant K (k_a/k_d) increases from blue, red, green and brown.

Derivation of kinetic parameters based on TPD results

Here we are going to show how to use the TPD technique to estimate desorption energy, reaction energy, as well as Arrhenius pre-exponential factor. Let us assume that molecules are irreversibly adsorbed on the surface at some low temperature T_0. The leak valve is closed, the valve to the pump is opened, and the density of product molecules is monitored with a mass spectrometer as the crystal is heated under programmed temperature,

$$T = T_0 + \beta T$$

where β is the heating rate ($\sim$10 °C/s). We know the desorption rate depends strongly on temperature, so when the temperature of the crystal reaches a high enough value so that the desorption rate is appreciable, the mass spectrometer will begin to record a rise in density. At higher temperatures, the surface will finally become depleted of desorbing molecules; therefore, the mass spectrometer signal will decrease. According to the shape and position of the peak in the mass signal, we can learn about the activation energy for desorption and the Arrhenius pre-exponential factor.

First-order process

Consider a first-order desorption process, with a rate constant k_d,

$$k_d = A\exp(-\Delta E_a/RT)$$

where A is Arrhenius pre- exponential factor. If θ is assumed to be the number of surface adsorbates per unit area, the desorption rate will be given by:

$$-d\theta/dt = k_d\theta = \theta A\exp(-\Delta E_a/RT)$$

Since we know the relationship between heat rate β and temperature on the crystal surface T,

$$T = T_0 + \beta t$$

$$1/dt = \beta/dT$$

Multiplying by $d\theta$ gives,

$$-d\theta/dt = -\beta(d\theta/dT)$$

since:

$$-d\theta/dt = k_d\theta = \theta A\exp(-\Delta E_a/RT)$$

$$-d\theta/dT = \theta A/\beta = (\theta A/\beta)\exp(-\Delta E_a/RT)$$

A plot of the form of $d\theta/dT$ versus T is shown in Figure 11.5.

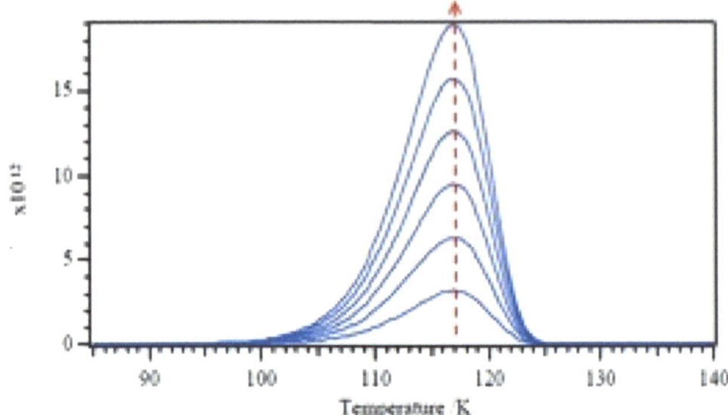

Figure 11.5: A simulated TPD experiment: Consider a first order reaction between adsorbates and surface. Values of T_m keep constant as the initial coverage θ from 1.0×10^{13} - 6.0×10^{13} cm^{-2}. $E_a = 30$ KJ/mol; $\beta = 1.5$ °C/s; $A = 1 \times 10^{13}$.

We notice that the T_m (peak maximum) in Figure 11.5 keeps constant with increasing θ, which means the value of T_m does not depend on the initial coverage θ in the first-order desorption. It is seen that the T_m values will increase with increasing E_a.

At the peak of the mass signal, the increase in the desorption rate is matched by the decrease in surface concentration per unit area so that the change in $d\theta/dT$ with T is zero:

$$-d\theta/dT = (\theta A/\beta)\exp(-\Delta E_a/RT)$$

$$(d/dT)[(\theta A/\beta)\exp(-\Delta E_a/RT)] = 0$$

$$(\Delta E_a/RT_m^2) = (1/\theta)(d\theta/dT)$$

then,

$$(\Delta E_a/RT_m^2) = (A/\beta)\exp(-\Delta E_a/RT_m)$$

$$2\ln T_m = -\ln\beta = (\Delta E_a/RT_M) + \ln(\Delta E_a/RA)$$

This tells us if different heating rates β are used and the left-hand side of the above equation is plotted as a function of $1/T_m$, we can see that a straight line should be obtained whose slope is $\Delta E_a/R$ and intercept is $\ln(\Delta E_a/RA)$. So, we are able to obtain the activation energy to desorption ΔE_a and Arrhenius pre-exponential factor A.

Second-order process

Now let consider a second-order desorption process,

$$2\text{A-S} \rightarrow \text{A}_2 + \text{S}$$

with a rate constant k_d, the desorption kinetics can be deduced as:

$$-d\theta/dT = \theta^2 A\exp(-\Delta E_a/RT)$$

The result is different from the first-order reaction whose T_m value does not depend upon the initial coverage the temperature of the peak T_m will decrease with increasing initial surface coverage (Figure 11.6).

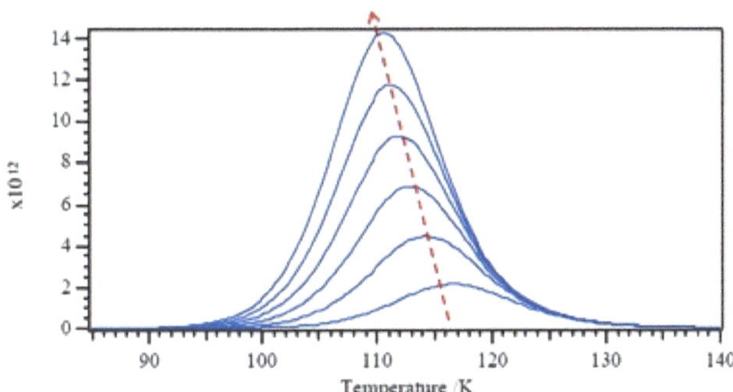

Figure 11.6: A simulated second order TPD experiment: A second-order reaction between adsorbates and surface. Values of T_m decrease as the initial coverage θ increases from 1.0×10^{13} - 6.0×10^{13} cm^{-2}. E_a = 30 KJ/mol; β = 1.5 °C/s; A = 1×10^{-1}.

Zero-order process

The zero-order desorption kinetics relationship as:

$$-d\theta/dT = A\exp(-\Delta E_a/RT)$$

Looking at desorption rate for the zero-order reaction (Figure 11.7), we can observe that the desorption rate does not depend on coverage and also implies that desorption rate increases exponentially with T. Also, according to the plot of desorption rate versus T, we figure out the desorption rate rapid drop when all molecules have desorbed. Plus, temperature of peak, T_m, moves to higher T with increasing coverage θ.

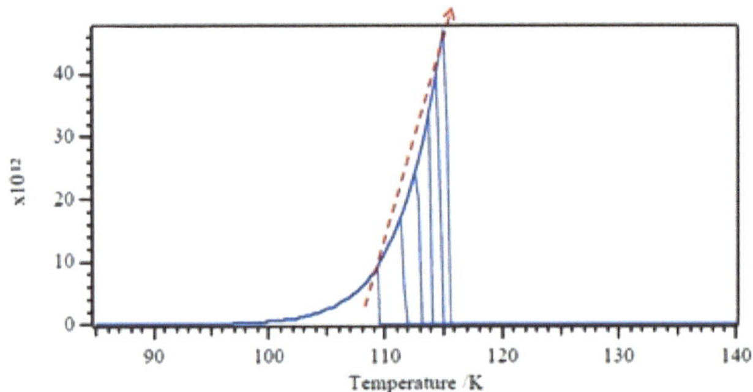

Figure 11.7: A simulated zero-order TPD experiment: A zero-order reaction between adsorbates and surface. Values of T_m increase apparently as the initial coverage θ increases from 1.0×10^{13} - 6.0×10^{13} cm^{-2}. $E_a = 30$ KJ/mol; $\beta = 1.5$ °C/s; $A = 1\times10^{28}$.

A typical example

A typical TPD spectra of deuterium (D_2) from Rh(100) for different exposures in Langmuirs ($L = 10^{-6}$ Torr-sec) shows in Figure 11.8. First, the desorption peaks from 08 – 20 exposure show two different desorbing regions. The higher one can undoubtedly be ascribed to chemisorbed D_2 on Rh(100) surface, which means chemisorbed molecules need higher energy used to overcome their activation energy for desorption. The lower desorption region is then due to physisorbed D_2 with much lower desorption activation energy than chemisorbed D_2. According to the TPD theory, the peak maximum shifts to lower temperature with increasing initial coverage, which means it should belong to a second-order reaction.

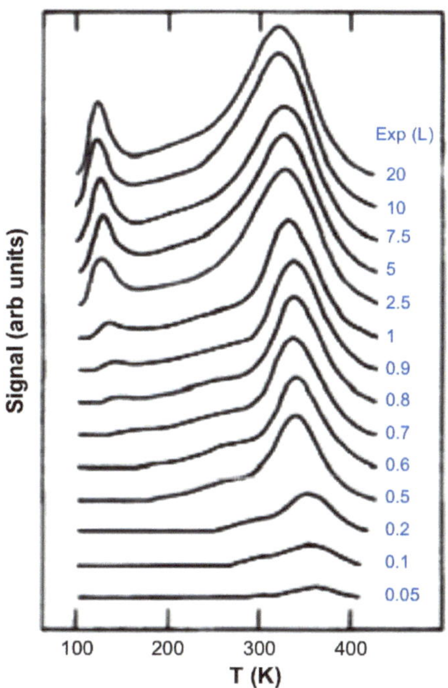

Figure 11.8: TPD spectra of D_2 from Rh(100) for different exposures in L (1 Langmuir = 10^{-6} Torr exposure in 1 s). Adapted from Y. Kim, H. C. Peebles, and J. M. White, Adsorption of D_2, CO and the interaction of CO-adsorbed D_2 and CO on Rh(100). *Surface Sci.***, 1982, 114, 363. Copyright: Elsevier (1982).**

Bibliography

J. L. Falconer and J. A. Schwarz, Temperature-programmed desorption and reaction: applications to supported catalysts. *Catal. Rev. Sci. Eng.*, 1983, **25**, 141.

C. V. Hidalgo, H. Itoh, T. Hattori, M. Niwa, and Y. Murakami, Measurement of the acidity of various zeolites by temperature-programmed desorption of ammonia. *J. Catal.*, 1984, **85**, 362.

P. L. Houston, *Chemical Kinetics and Reaction Dynamics*, McGraw Hill, New York (2001).

Y. Kim, H. C. Peebles, and J. M. White, Adsorption of D_2, CO and the interaction of CO-adsorbed D_2 and CO on Rh(100). *Surface Sci.*, 1982, **114**, 363.

I. Langmuir, The constitution and fundamental properties of solids and liquids. Part I. solids. *J. Am. Chem. Soc.*, 1916, **38**, 221.

P. A. Redhead, Thermal desorption of gases. *Vacuum*, 1962, **12**, 203.

Chapter 12: Magnetism

Samuel Maguire-Boyle and Andrew R. Barron

Magnetics

Magnetic moments

The magnetic moment of a material is the incomplete cancellation of the atomic magnetic moments in that material. Electron spin and orbital motion both have magnetic moments associated with them (Figure 12.1), but in most atoms the electronic moments are oriented usually randomly so that overall in the material they cancel each other out (Figure 12.2), this is called diamagnetism.

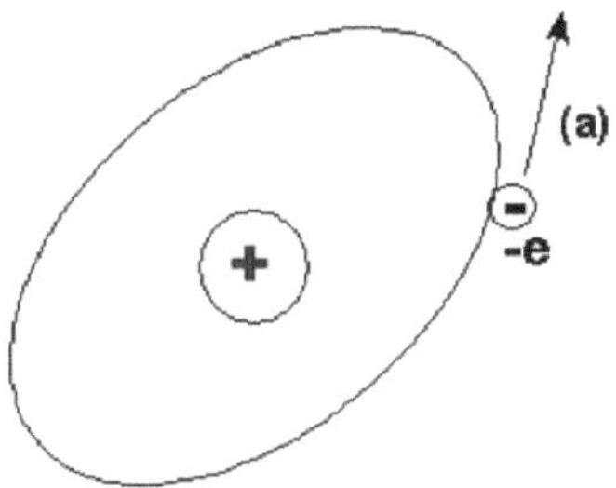

Figure 12.1: Orbital magnetic moment.

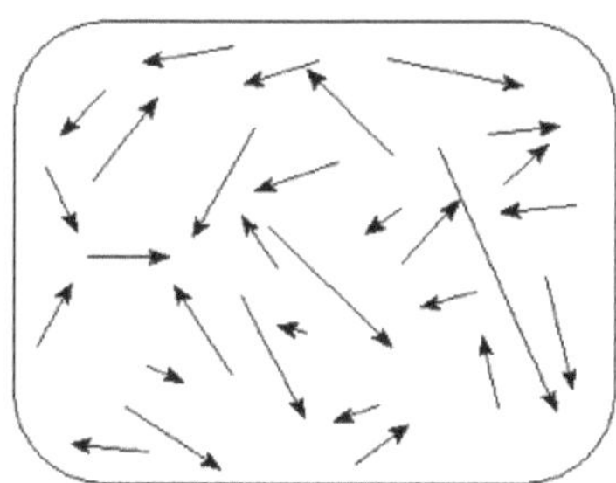

Figure 12.2: Magnetic moments in a diamagnetic sample.

If the cancellation of the moments is incomplete, then the atom has a net magnetic moment. There are multiple subclasses of magnetic ordering, such as para-, superpara-, ferro-, antiferro- or ferro-magnetism which can be displayed in a material and that usually depends, upon the strength and type of magnetic interactions and external parameters such as temperature and crystal structure atomic content and the magnetic environment which a material is placed in. The magnetic moments of atoms, molecules or formula units are often quoted in terms of the Bohr magneton, which is equal to the magnetic moment due to electron spin,

$$\mu_B = \frac{e\hbar}{2m_e c} = 9.72\times10^{-24} \text{ J/T}$$

where e is the elementary charge, $\hbar$ is the reduced Planck constant, m_e is the electron rest mass, and c is the speed of light.

Magnetization

The magnetism of a material, the extent that which a material is magnetic, is not a static quantity, but varies compared to the environment that a material is placed in. It is similar to the temperature of a material. For example, if a material is placed in an oven it will heat up to a temperature similar to that of the ovens. However, the speed of heating of that material, and also that of cooling are determined by the atomic structure of the material. The magnetization of a material is similar. When a material is placed in a magnetic field it maybe become magnetized to an extent and retain that magnetization after it is removed from the field. The extent of magnetization, and type of magnetization and the length of time that a material remains magnetized, depends again on the atomic makeup of the material.

Measuring a materials magnetism can be done on a micro or macro scale. Magnetism is measured over two parameters direction and strength. Thus, magnetization has a vector quantity. The simplest form of a magnetometer is a compass. It measures the direction of a magnetic field. However, more sophisticated instruments have been developed which give a greater insight into a materials magnetism.

So, what exactly are you reading when you observe the output from a magnetometer? The magnetism of a sample is called the magnetic moment of that sample and will be called that from now on. The single value of magnetic moment for the sample, is a combination of the magnetic moments on the atoms within the sample (Figure 12.3), it is also the type and level of magnetic ordering and the physical dimensions of the sample itself.

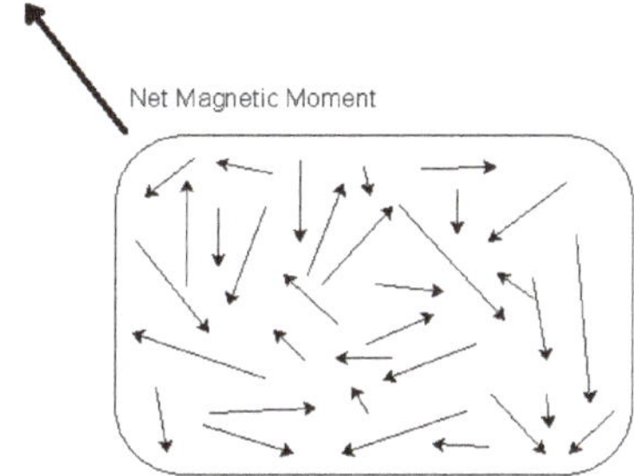

Figure 12.3: Schematic representations of the net magnetic moment in a diamagnetic sample.

The "intensity of magnetization", M, is a measure of the magnetization of a body. It is defined as the magnetic moment per unit volume,

$$M = m/V$$

with units of Am (emu.cm^3 in cgs notation).

A material contains many atoms and their arrangement affects the magnetization of that material. In Figure 12.4 a magnetic moment, m, is contained in unit volume. This has a magnetization of m Am. Figure 12.4b shows two such units, with the moments aligned parallel. The vector sum of moments is 2m in this case, but as the both the moment and volume are doubled M remains the same. In Figure 12.4c the moments are aligned antiparallel. The vector sum of moments is now 0 and hence the magnetization is 0 Am.

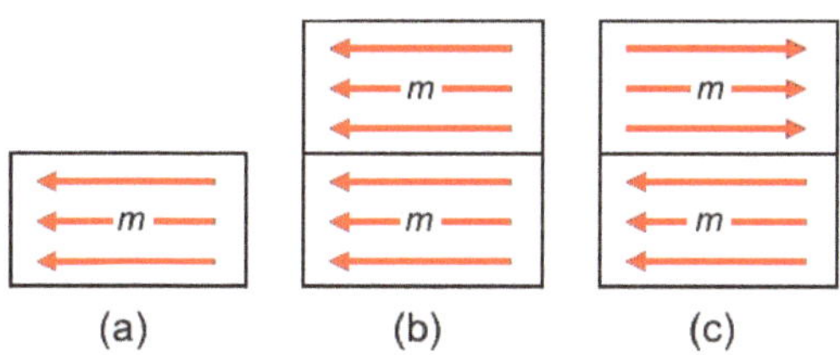

Figure 12.4: Effect of moment alignment on magnetization: (a) Single magnetic moment, (b) two identical moments aligned parallel and (c) antiparallel to each other. Adapted from J. Bland, Thesis M. Phys (Hons)., *A Mossbauer Spectroscopy and Magnetometry Study of Magnetic Multilayers and Oxides.* **Oliver Lodge Labs, Department of Physics, University of Liverpool (2002).**

Scenarios (b) and (c) are a simple representation of ferro- and antiferromagnetic ordering. Hence, we would expect a large magnetization in a ferromagnetic material such as pure iron and a small magnetization in an antiferromagnet such as γ-Fe_2O_3.

Magnetic response

When a material is passed through a magnetic field it is affected in two ways:
- through its susceptibility,
- through its permeability.

Magnetic susceptibility

The concept of magnetic moment is the starting point when discussing the behavior of magnetic materials within a field. If you place a bar magnet in a field, it will experience a torque or moment tending to align its axis in the direction of the field (Figure 12.5). A compass needle behaves in the same way. This torque increases with the strength of the poles and their distance apart. So, the value of magnetic moment tells you, in effect, 'how big a magnet' you have.

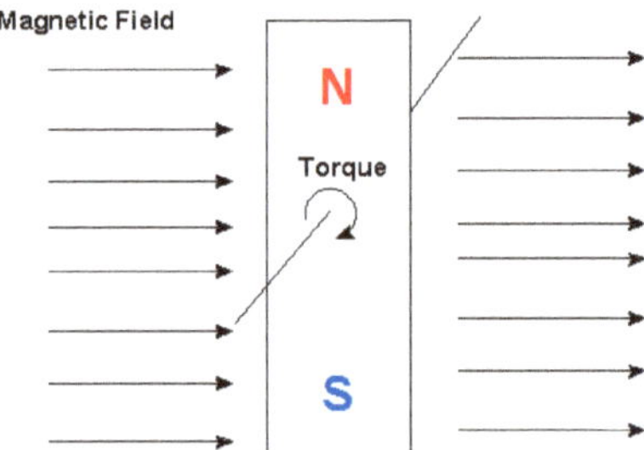

Figure 12.5: Schematic representation of the torque or moment that a magnet experiences when it is placed in a magnetic field. The magnetic will try to align with the magnetic field.

If you place a material in a weak magnetic field, the magnetic field may not overcome the binding energies that keep the material in a non-magnetic state. This is because it is energetically more favorable for the material to stay exactly the same. However, if the strength of the magnetic moment is increased, the torque acting on the smaller moments in the material, it may become energetically more preferable for the material to become magnetic. The reasons that the material becomes magnetic depends on factors such as crystal structure the temperature of the material and the strength of the field that it is in. However, a simple explanation of this is that as the magnetic moment strength increases it becomes more favorable for the small fields to align themselves along the path of the magnetic field, instead of being opposed to the system. For this to occur the material must rearrange its magnetic makeup at the atomic level to lower the energy of the system and restore a balance.

It is important to remember that when we consider the magnetic susceptibility and take into account how a material changes on the atomic level when it is placed in a magnetic field with a certain moment. The moment that we are measuring with our magnetometer is the total moment of that sample.

$$\chi = M/H$$

where χ = susceptibility, M = variation of magnetization, and H = applied field.

Magnetic permeability

Magnetic permeability is the ability of a material to conduct an electric field. In the same way that materials conduct or resist electricity, materials also conduct or resist a magnetic flux or the flow of magnetic lines of force (Figure 12.6).

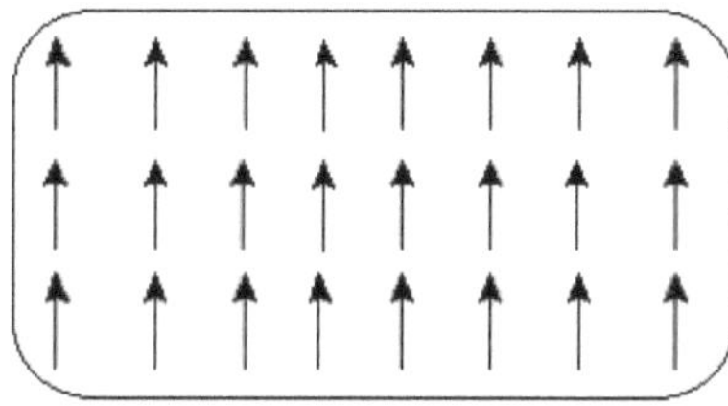

Figure 12.6: Magnetic ordering in a ferromagnetic material.

Ferromagnetic materials are usually highly permeable to magnetic fields. Just as electrical conductivity is defined as the ratio of the current density to the electric field strength, so the magnetic permeability, μ, of a particular material is defined as the ratio of flux density to magnetic field strength,

$$\mu = B/H$$

However unlike in electrical conductivity magnetic permeability for a ferromagnetic material is nonlinear. Permeability, where μ is written without a subscript, is known as absolute permeability. Instead a variant is used called relative permeability. Relative permeability is a variation upon 'straight' or absolute permeability, μ, but is more useful as it makes clearer how the presence of a particular material affects the relationship between flux density and field strength. The term 'relative' arises because this permeability is defined in relation to the permeability of a vacuum, μ_0.

$$\mu_r = \mu/\mu_o$$

For example, if you use a material for, which $\mu_r = 3$, then you know that the flux density will be three times as great as it would be if we just applied the same field strength to a vacuum.

Initial permeability describes the relative permeability of a material at low values of B (below 0.1 T). The maximum value for μ in a material is frequently a factor of between 2 and 5 or more above its initial value. Low flux has the advantage that every ferrite can be measured at that density without risk of saturation. This consistency means that comparison between different ferrites is easy. Also, if you measure the inductance with a normal component bridge then you are doing so with respect to the initial permeability.

The permeability of a vacuum has a finite value, about 1.257×10^{-6} H/m, and is denoted by the symbol μ_0. Note that this value is constant with field strength and temperature. Contrast this with the situation in ferromagnetic materials where μ is strongly dependent upon both. Also, for practical purposes, most non-ferromagnetic substances (such as wood, plastic, glass, bone, copper aluminum, air and water) have permeability almost equal to μ_0; that is, their relative permeability is 1.0.

The permeability, μ, the variation of magnetic induction,

$$B = \mu_0(H+M)$$

with applied field,

$$\mu = B/H$$

Background contributions

A single measurement of a sample's magnetization is relatively easy to obtain, especially with modern technology. Often it is simply a case of loading the sample into the magnetometer in the correct manner and performing a single measurement. This value is, however, the sum total of the sample, any substrate or backing and the sample mount. A sample substrate can produce a substantial contribution to the sample total.

For substrates that are diamagnetic, under zero applied field, this means it has no effect on the measurement of magnetization. Under applied fields its contribution is linear and temperature independent. The diamagnetic contribution can be calculated from knowledge of the volume and properties of the substrate and subtracted as a constant linear term to produce the signal from the sample alone. The diamagnetic background can also be seen clearly at high fields where the sample has reached saturation: the sample saturates but the linear background from the substrate continues to increase with field. The gradient of this background can be recorded and subtracted from the readings if the substrate properties are not known accurately.

Hysteresis

When a material exhibits hysteresis, it means that the material responds to a force and has a history of that force contained within it. Consider if you press on something until it depresses. When you release that pressure, if the material remains depressed and doesn't spring back then it is said to exhibit some type of hysteresis. It remembers a history of what happened to it and may exhibit that history in some way. Consider a piece of iron that is brought into a magnetic field, it retains some magnetization, even after the external magnetic field is removed. Once magnetized, the iron will stay magnetized indefinitely. To demagnetize the iron, it is necessary to apply a magnetic field in the opposite direction. This is the basis of memory in a hard disk drive.

The response of a material to an applied field and its magnetic hysteresis is an essential tool of magnetometry. Paramagnetic and diamagnetic materials can easily be recognized, soft and hard ferromagnetic materials give different types of hysteresis curves and from these curve values such as saturation magnetization, remnant magnetization and coercivity are readily observed. More detailed curves can give indications of the type of magnetic interactions within the sample.

Diamagnetism and paramagnetism

The intensity of magnetization depends upon both the magnetic moments in the sample and the way that they are oriented with respect to each other, known as the magnetic ordering.

Diamagnetic materials, which have no atomic magnetic moments, have no magnetization in zero field. When a field is applied a small, negative moment is induced on the diamagnetic atoms proportional to the applied field strength. As the field is reduced the induced moment is reduced (Figure 12.7).

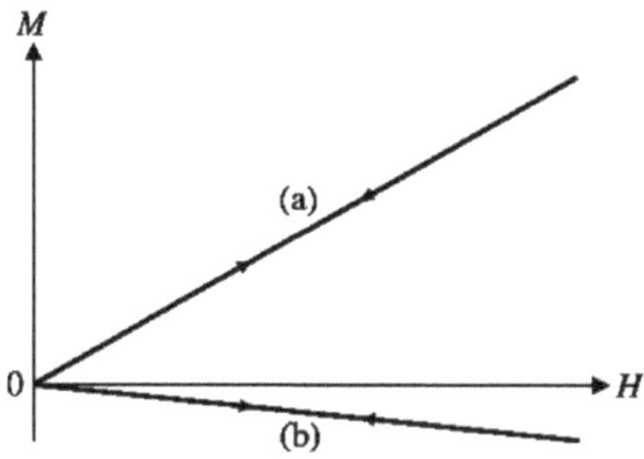

Figure 12.7: Typical effect on the magnetization, M, of an applied magnetic field, H, on (a) a para- magnetic system and (b) a diamagnetic system. Adapted from J. Bland, Thesis M. Phys (Hons)., *A Mossbauer Spectroscopy and Magnetometry Study of Magnetic Multilayers and Oxides.* **Oliver Lodge Labs, Department of Physics, University of Liverpool (2002).**

In a paramagnet the atoms have a net magnetic moment but are oriented randomly throughout the sample due to thermal agitation, giving zero magnetization. As a field is applied the moments tend towards alignment along the field, giving a net magnetization which increases with applied field as the moments become more ordered. As the field is reduced the moments become disordered again by their thermal agitation. Figure 12.7 shows the linear response M v H where $\mu H \ll kT$.

Ferromagnetism

The hysteresis curves for a ferromagnetic material are more complex than those for diamagnets or paramagnets. Figure 12.8 shows the main features of such a curve for a simple ferromagnet.

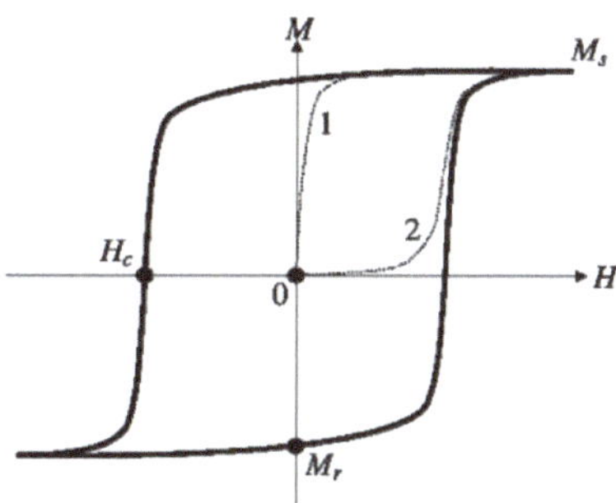

Figure 12.8: Schematic of a magnetization hysteresis loop in a ferromagnetic material showing the saturation magnetization, M_s, coercive field, H_c, and remnant magnetization, M_r. Virgin curves are shown dashed for nucleation (1) and pinning (2) type magnets. Adapted from J. Bland, Thesis M. Phys (Hons)., *A Mossbauer Spectroscopy and Magnetometry Study of Magnetic Multilayers and Oxides*. Oliver Lodge Labs, Department of Physics, University of Liverpool (2002).

In the virgin material (point 0) there is no magnetization. The process of magnetization, leading from point 0 to saturation at $M = M_s$, is outlined below. Although the material is ordered ferromagnetically it consists of a number of ordered domains arranged randomly giving no net magnetization. This is shown in Figure 12.9a with two domains whose individual saturation moments, M_s, lie antiparallel to each other. As the magnetic field, H, is applied, Figure 12.9b, those domains which are more energetically favorable increase in size at the expense of those whose moment lies more antiparallel to H. There is now a net magnetization; M. Eventually a field is reached where all of the material is a single domain with a moment aligned parallel, or close to parallel, with H.

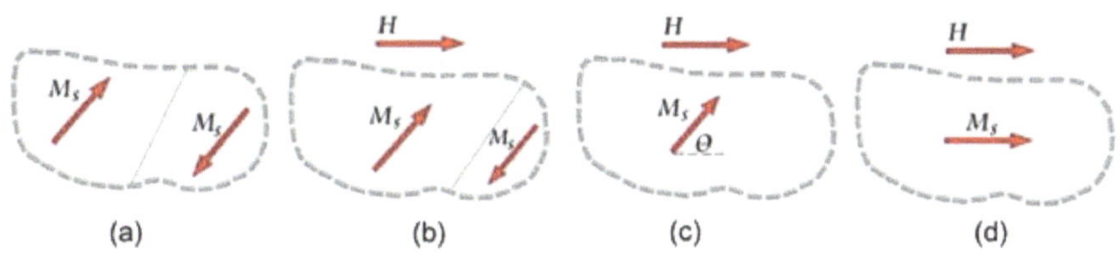

Figure 12.9: The process of magnetization in a demagnetized ferromagnet. Adapted from J. Bland, Thesis M. Phys (Hons)., *A Mossbauer Spectroscopy and Magnetometry Study of Magnetic Multilayers and Oxides*. Oliver Lodge Labs, Department of Physics, University of Liverpool (2002).

The magnetization is now,

$$M = M_s cos\Theta$$

where Θ is the angle between M_S along the easy magnetic axis and H. Finally, M_S is rotated parallel to H and the ferromagnet is saturated with a magnetization $M = M_s$. The process of domain wall motion affects the shape of the virgin curve. There are two qualitatively different modes of behavior known as nucleation and pinning, shown in Figure 12.8 as curves 1 and 2, respectively.

In a nucleation-type magnet saturation is reached quickly at a field much lower than the coercive field. This shows that the domain walls are easily moved and are not pinned significantly. Once the domain structure has been removed the formation of reversed domains becomes difficult, giving high coercivity. In a pinning-type magnet fields close to the coercive field are necessary to reach saturation magnetization. Here the domain walls are substantially pinned, and this mechanism also gives high coercivity.

Remnance

As the applied field is reduced to 0 after the sample has reached saturation the sample can still possess a remnant magnetization, M_r. The magnitude of this remnant magnetization is a product of the saturation magnetization, the number and orientation of easy axes and the type of anisotropy symmetry. If the axis of anisotropy or magnetic easy axis is perfectly aligned with the field then $M_r = M_s$, and if perpendicular, $M_r = 0$.

At saturation the angular distribution of domain magnetizations is closely aligned to H. As the field is removed, they turn to the nearest easy magnetic axis. In a cubic crystal with a positive anisotropy constant, K_1, the easy directions are <100>. At remnance the domain magnetizations will lie along one of the three <100> directions. The maximum deviation from H occurs when H is along the <111> axis, giving a cone of distribution of 55° around the axis. Averaging the saturation magnetization over this angle gives a remnant magnetization of 0.832 Ms.

Coercivity

The coercive field, H_c, is the field at which the remnant magnetization is reduced to zero. This can vary from a few Am for soft magnets to 10^7 Am for hard magnets. It is the point of magnetization reversal in the sample, where the barrier between the two states of magnetization is reduced to zero by the applied field allowing the system to make a Barkhausen jump to a lower energy. It is a general indicator of the energy gradients in the sample which oppose large changes of magnetization.

The reversal of magnetization can come about as a rotation of the magnetization in a large volume or through the movement of domain walls under the pressure of the applied field (Figure 12.10). In general materials with few or no domains have a high coercivity whilst those with many domains have a low coercivity. However, domain wall pinning by physical defects such as vacancies, dislocations and grain boundaries can increase the coercivity.

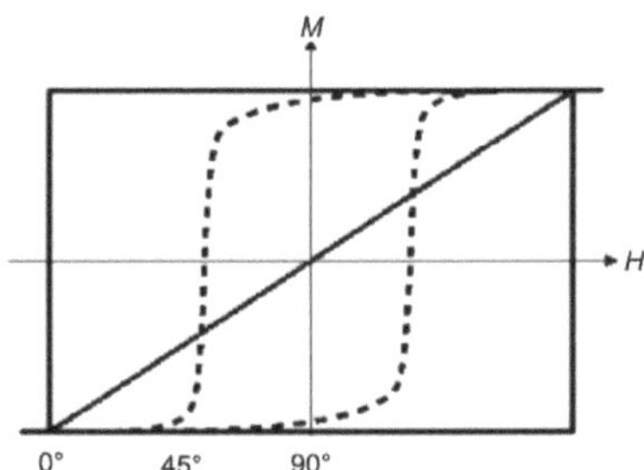

Figure 12.10: Shape of hysteresis loop as a function of Θ_H, the angle between anisotropy axis and applied field H, for Θ_H, = 0°, 45°, and 90°. Adapted from J. Bland, Thesis M. Phys (Hons)., *A Mossbauer Spectroscopy and Magnetometry Study of Magnetic Multilayers and Oxides*. Oliver Lodge Labs, Department of Physics, University of Liverpool (2002).

The loop illustrated in Figure 12.10 is indicative of a simple bi-stable system. There are two energy minima: one with magnetization in the positive direction, and another in the negative direction. The depth of these minima is influenced by the material and its geometry and is a further parameter in the strength of the coercive field. Another is the angle, Θ_H, between the anisotropy axis and the applied field. The above g shows how the shape of the hysteresis loop and the magnitude of H_c varies with Θ_H. This effect shows the importance of how samples with strong anisotropy are mounted in a magnetometer when comparing loops.

Temperature dependence

A hysteresis curve gives information about a magnetic system by varying the applied field, but important information can also be gleaned by varying the temperature. As well as indicating transition temperatures, all of the main groups of magnetic ordering have characteristic temperature/magnetization curves. These are summarized in Figures 12.11 and 12.12. At all temperatures a diamagnet displays only any magnetization induced by the applied field and a small, negative susceptibility. The curve shown for a paramagnet (Figure 12.11) is for one obeying the Curie law,

$$\chi = C/T$$

and so, intercepts the axis at $T = 0$. This is a subset of the Curie-Weiss law,

$$\chi = C/(T-\theta)$$

where θ is a specific temperature for a particular substance (equal to 0 for paramagnets).

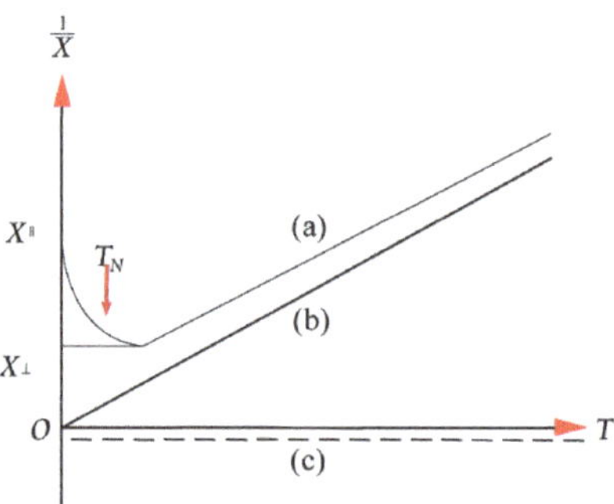

Figure 12.11: Variation of reciprocal susceptibility with temperature for: (a) antiferromagnetic, (b) paramagnetic and (c) diamagnetic ordering. Adapted from J. Bland, Thesis M. Phys (Hons), *A Mossbauer Spectroscopy and Magnetometry Study of Magnetic Multilayers and Oxides*. Oliver Lodge Labs, Department of Physics, University of Liverpool (2002).

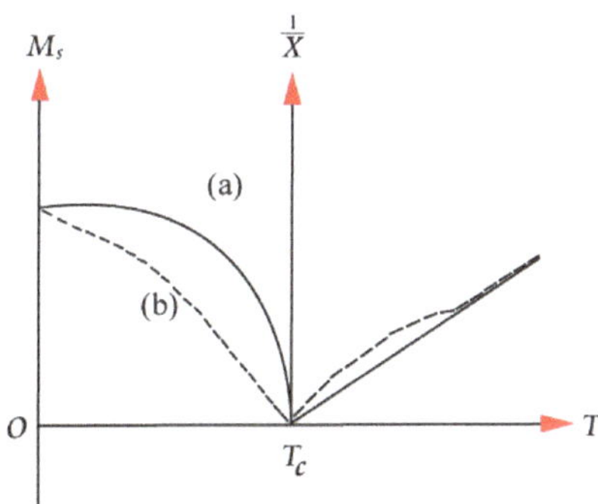

Figure 12.12: Variation of saturation magnetization below, and reciprocal susceptibility above T_c for: (a) ferromagnetic and (b) ferrimagnetic ordering. Adapted from J. Bland, Thesis M. Phys (Hons), *A Mossbauer Spectroscopy and Magnetometry Study of Magnetic Multilayers and Oxides*. Oliver Lodge Labs, Department of Physics, University of Liverpool (2002).

Above T_N and T_c both antiferromagnets and ferromagnets behave as paramagnets with $^1/_\chi$ linearly proportional to temperature. They can be distinguished by their intercept on the temperature axis, $T = \Theta$. Ferromagnetics have a large, positive Θ, indicative of their strong interactions. For paramagnetics $\Theta = 0$ and antiferromagnetics have a negative Θ.

The net magnetic moment per atom can be calculated from the gradient of the straight-line graph of $^1/_\chi$ versus temperature for a paramagnetic ion, rearranging Curie's law to give,

$$\mu = \sqrt{(3Ak/N_x)}$$

where A is the atomic mass, k is Boltzmann's constant, N is the number of atoms per unit volume and x is the gradient.

Ferromagnets below T_C display spontaneous magnetization. Their susceptibility above T_C in the para- magnetic region is given by the Curie-Weiss law

where g is the gyromagnetic constant. In the ferromagnetic phase with T greater than T_C the magnetization M (T) can be simplified to a power law, for example the magnetization as a function of temperature can be given by

$$M(T) \approx (T_c - T)^\beta$$

where the term β is typically in the region of 0.33 for magnetic ordering in three dimensions.

The susceptibility of an antiferromagnet increases to a maximum at T_N as temperature is reduced, then decreases again below T_N. In the presence of crystal anisotropy in the system this change in susceptibility depends on the orientation of the spin axes: χ (parallel) decreases with temperature whilst χ (perpendicular) is constant. These can be expressed as

$$\chi_{\perp} = C/2\theta$$

where C is the Curie constant and Θ is the total change in angle of the two sublattice magnetizations away from the spin axis, and

$$\chi_{||} = \frac{2\eta_g\mu_H^2 B'(J,a_o')}{2kT + \eta_g\mu_H^2\gamma\rho B'(J,a_o')}$$

where ng is the number of magnetic atoms per gram, B' is the derivative of the Brillouin function with respect to its argument a', evaluated at a_o', μ_H is the magnetic moment per atom and γ is the molecular field coefficient.

Theory of a superconducting quantum interference device (SQUID)

Introduction

One of the most sensitive forms of magnetometry is SQUID magnetometry. This uses technique uses a com- bination of superconducting materials and Josephson junctions to measure magnetic fields with resolutions up to $\sim 10^{-14}$ kG or greater. In the proceeding pages we will describe how a SQUID actually works.

Electron-pair waves

In superconductors the resistanceless current is carried by pairs of electrons, known as Cooper Pairs. A Cooper Pair is a pair of electrons. Each electron has a quantized wavelength. With a Cooper pair each electron wave couples with its opposite number over a large distance. This phenomenon is a result of the very low temperatures at which many materials will superconduct.

What exactly is superconductance? When a material is at very low temperatures, its crystal lattice behaves differently than when it is at higher temperatures. Usually at higher temperatures a material will have large vibrations called in the crystal lattice. These vibrations scatter electrons as they pass through this lattice (Figure 12.13), and this is the basis for bad conductance.

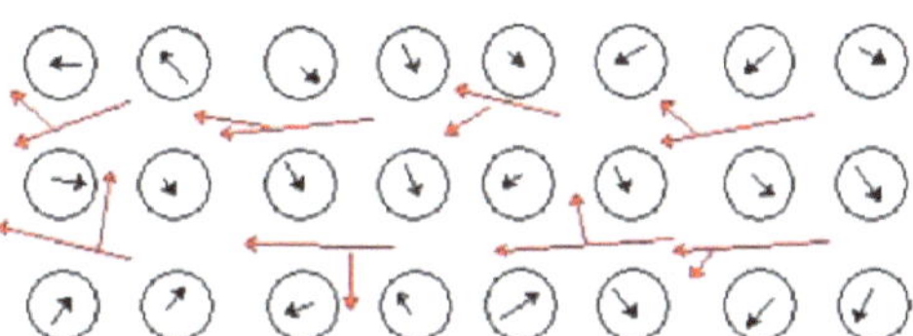

Figure 12.13: Schematic representation of the scattering of electrons as they pass through a vibrating lattice.

With a superconductor the material is designed to have very small vibrations, these vibrations are lessened even more by cooling the material to extremely low temperatures. With no vibrations there is no scattering of the electrons and this allows the material to superconduct.

The origin of a Cooper pair is that as the electron passes through a crystal lattice at superconducting temperatures its negative charge pulls on the positive charge of the nuclei in the lattice through coulombic interactions producing a ripple. An electron traveling in the opposite direction is attracted by this ripple. This is the origin of the coupling in a Cooper pair (Figure 12.14).

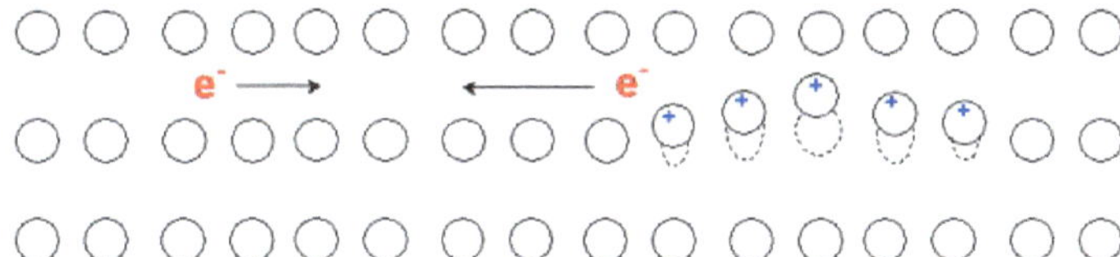

Figure 12.14: Schematic representation of the Cooper pair coupling model.

A passing electron attracts the lattice, causing a slight ripple toward its path. Another electron passing in the opposite direction is attracted to that displacement (Figure 12.15).

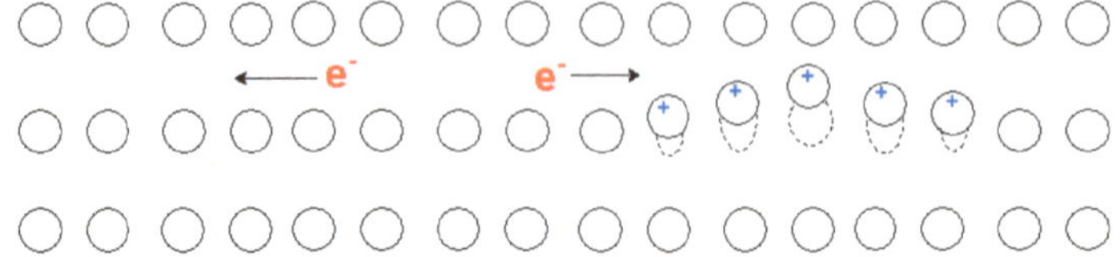

Figure 12.15: Schematic representation of Cooper pair coupling

Due to the coupling and the fact that for each pair there is two spin states (Figure 12.16).

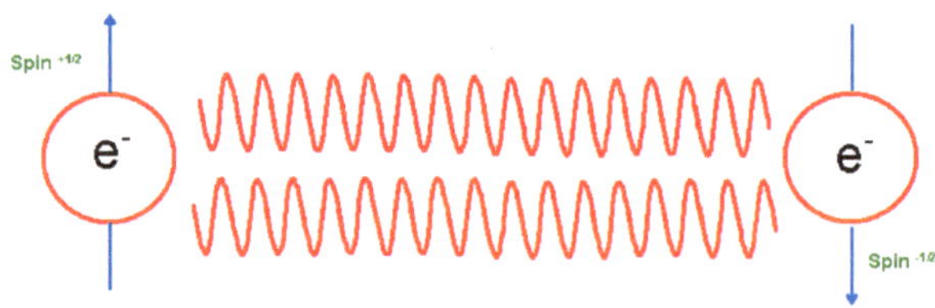

Figure 12.16: Schematic representation of the condensation of the wavelengths of a Cooper pairs

Each pair can be treated as a single particle with a whole spin, not half a spin such as is usually the case with electrons. This is important, as an electron which is classed in a group of matter called Fermions are governed by the Fermi exclusion principle which states that anything with a spin of one half cannot occupy the same space as something with the same spin of one half. This turns the electron means that a Cooper pair is in fact a Boson the opposite of a Fermion and this allows the Coopers pairs to condensate into one wave packet. Each Coopers pair has a mass and charge twice that of a single electron, whose velocity is that of the center of mass of the pair. This coupling can only happen in extremely cold conditions as thermal vibrations become greater than the force that an electron can exert on a lattice, and thus, scattering occurs.

Each pair can be represented by a wavefunction of the form

$$\Phi_p = \Phi e^{i(P,r)/\hbar}$$

where P is the net momentum of the pair whose center of mass is at r. However, all the Cooper pairs in a superconductor can be described by a single wavefunction yet again due to the fact that the electrons are in a Coopers pair state and are thus Bosons in the absence of a current because all the pairs have the same phase - they are said to be "phase coherent",

$$\Psi_p = \Psi e^{i(P,r)/\hbar}$$

This electron-pair wave retains its phase coherence over long distances, and essentially produces a standing wave over the device circuit. In a SQUID there are two paths which form a circle and are made with the same standing wave (Figure 12.17). The wave is split in two sent off along different paths, and then recombined to record an interference pattern by adding the difference between the two.

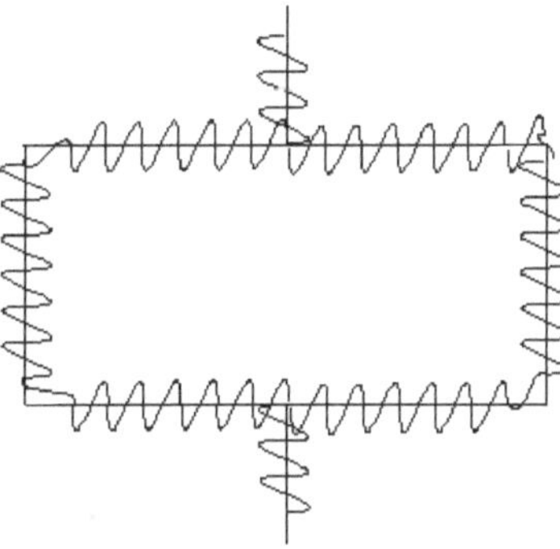

Figure 12.17: Schematic representation of a standing wave across a SQUID circuit.

This allows measurement at any phase differences between the two components, which if there is no interference will be exactly the same, but if there is a difference in their path lengths or in some interaction that the waves encounters such as a magnetic field it will correspond in a phase difference at the end of each path length.

A good example to use is of two water waves emanating from the same point. They will stay in phase if they travel the same distance but will fall out of phase if one of them has to deviate around an obstruction such as a rock. Measuring the phase difference between the two waves then provides information about the obstruction.

Phase and coherence

Another implication of this long-range coherence is the ability to calculate phase and amplitude at any point on the wave's path from the knowledge of its phase and amplitude at any single point, combined with its wavelength and frequency. The wavefunction of the electron-pair wave in the above eqn. can be rewritten in the form of a one-dimensional wave as

$$\Psi_p = \Psi \sin 2\pi[(x/\lambda)-vt]$$

If we take the wave frequency, V, as being related to the kinetic energy of the Cooper pair with a wavelength, λ, being related to the momentum of the pair by the relation $\lambda = h/p$ then it is possible to evaluate the phase difference between two points in a current carrying superconductor. If a resistanceless current flows between points X and Y on a superconductor there will be a phase difference between these points that is constant in time.

Effect of a magnetic field

The parameters of a standing wave are dependent on a current passing through the circuit; they are also strongly affected by an applied magnetic field. In the presence of a magnetic field the momentum, p, of a particle with charge q in the presence of a magnetic field becomes $mV + qA$ where A is the magnetic vector potential. For electron-pairs in an applied field their moment P is now,

$$P = 2mV+2eA$$

In an applied magnetic field, the phase difference between points X and Y is now a combination of that due to the supercurrent and that due to the applied field.

The fluxoid

One effect of the long-range phase coherence is the quantization of magnetic flux in a superconducting ring. This can either be a ring, or a superconductor surrounding a non-superconducting region. Such an arrangement can be seen in Figure 12.18 where region N has a flux density B within it due to supercurrents flowing around it in the superconducting region S.

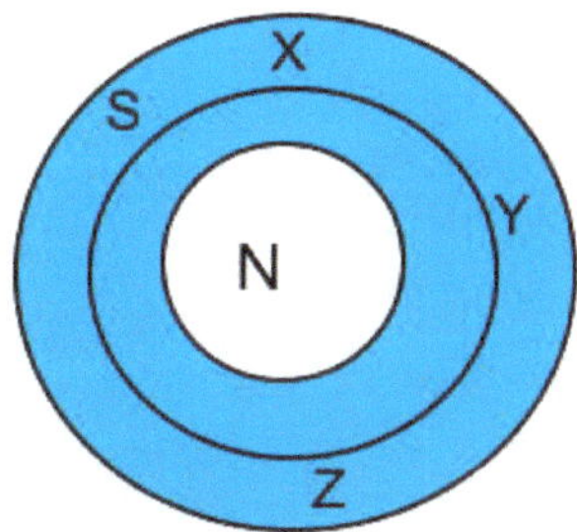

Figure 12.18: Superconductor enclosing a non-superconducting region. Adapted from J. Bland, Thesis M. Phys (Hons), *A Mossbauer Spectroscopy and Magnetometry Study of Magnetic Multilayers and Oxides.* **Oliver Lodge Labs, Department of Physics, University of Liverpool (2002).**

In the closed path XYZ encircling the non-superconducting region there will be a phase difference of the electron-pair wave between any two points, such as X and Y, on the curve due to the field and the circulating current.

If the superelectrons are represented by a single wave, then at any point on XYZX it can only have one value of phase and amplitude. Due to the long-range coherence the phase is single valued also called quantized meaning around the circumference of the ring $\Delta\Phi$ must equal $2\pi n$ where n is any integer. Due to the wave only having a single value the fluxoid can only exist in quantized units. This quantum is termed the fluxon, Φ_0, given by

$$\Phi_0 = h/2e = 2.07 \times 10^{-15} \text{ Wb}$$

Josephson tunneling

If two superconducting regions are kept totally isolated from each other the phases of the electron-pairs in the two regions will be unrelated. If the two regions are brought together then as they come close electron-pairs will be able to tunnel across the gap and the two electron-pair waves will become coupled. As the separation decreases, the strength of the coupling increases. The tunneling of the electron-pairs across the gap carries with it a superconducting current as predicted by B. D. Josephson (Figure 12.19) and is called "Josephson tunneling" with the junction between the two superconductors called a "Josephson junction" (Figure 12.20).

Figure 12.19: Welsh theoretical physicist Brian David Josephson, FRS (1940 -).

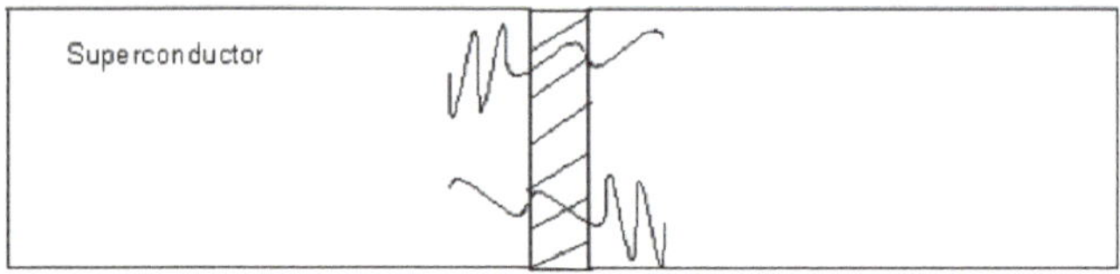

Figure 12.20: Schematic representation of the tunneling of Cooper pairs across a Josephson junction.

The Josephson tunneling junction is a special case of a more general type of weak link between two superconductors. Other forms include constrictions and point contacts, but the general form is of a region between two superconductors which has a much lower critical current and through which a magnetic field can penetrate.

Superconducting quantum interference device (SQUID)

A superconducting quantum interference device (SQUID) uses the properties of electron-pair wave coherence and Josephson Junctions to detect very small magnetic fields. The central element of a SQUID is a ring of superconducting material with one or more weak links called Josephson's junctions. An example is shown in Figure 12.21. With weak links at points W and X whose critical current, i_c, is much less than the critical current of the main ring. This produces a very low current density making the momentum of the electron-pairs small. The wavelength of the electron-pairs is thus very long leading to little difference in phase between any parts of the ring.

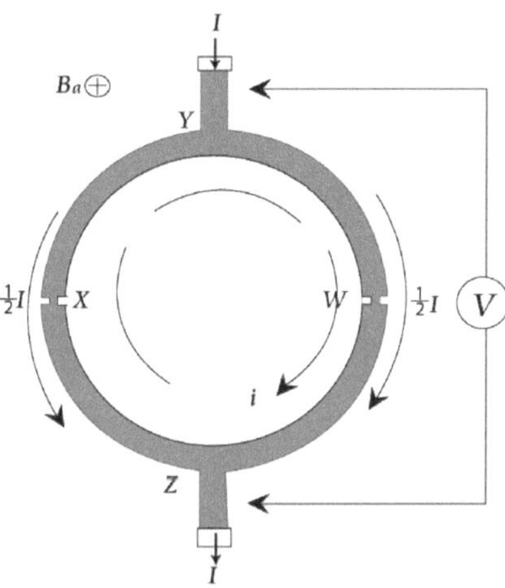

Figure 12.21: Superconducting quantum interference device (SQUID) as a simple magnetometer. Adapted from J. Bland, Thesis M. Phys (Hons), *A Mossbauer Spectroscopy and Magnetometry Study of Magnetic Multilayers and Oxides*. Oliver Lodge Labs, Department of Physics, University of Liverpool (2002).

If a magnetic field, B_a, is applied perpendicular to the plane of the ring (Figure 12.22), a phase difference is produced in the electron-pair wave along the path XYW and WZX. One of the features of a superconducting loop is that the magnetic flux, Φ, passing through it which is the product of the magnetic field and the area of the loop and is quantized in units of $\Phi_0 = h/(2e)$, where h is Planck's constant, 2e is the charge of the Cooper pair of electrons, and Φ has a value of 2×10^{15} tesla.m^2. If there are no obstacles in the loop, 0 then the superconducting current will compensate for the presence of an arbitrary magnetic field so that the total flux through the loop (due to the external field plus the field generated by the current) is a multiple of Φ_0.

Josephson predicted that a superconducting current can be sustained in the loop, even if its path is interrupted by an insulating barrier or a normal metal. The SQUID has two such barriers or Josephson junctions. Both junctions introduce the same phase difference when the magnetic flux through the loop is 0, Φ_0, $2\Phi_0$ and so on, which results in constructive interference, and they introduce opposite phase difference when the flux is $\Phi_0/2$, $3\Phi_0/2$ and so on, which leads to destructive interference. This interference causes the critical current density, which is the maximum current that the device can carry without dissipation, to vary. The critical current is so sensitive to the magnetic flux through the superconducting loop that even tiny magnetic moments can be measured. The critical current is usually obtained by measuring the voltage drop across the junction as a function of the total current through the device. Commercial SQUIDs transform the modulation in the critical current to a voltage modulation, which is much easier to measure.

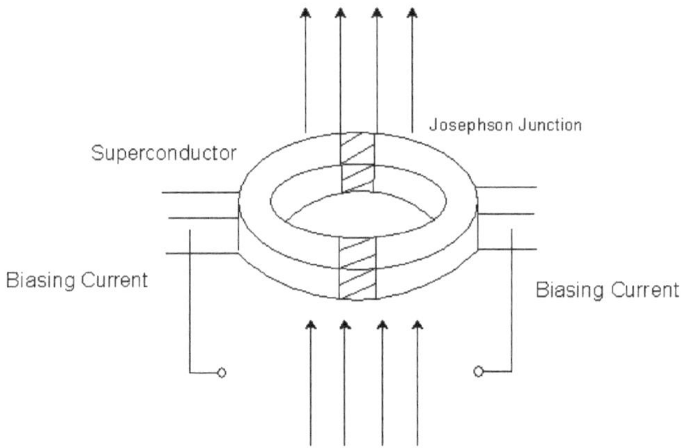

Figure 12.22: Schematic representation of a SQUID placed in a magnetic field.

An applied magnetic field produces a phase change around a ring, which in this case is equal,

$$\Delta\phi(B) = 2\pi(\Phi_a/\Phi_0)$$

where Φ_a is the flux produced in the ring by the applied magnetic field. The magnitude of the critical measuring current is dependent upon the critical current of the weak-links and the limit of the phase change around the ring being an integral multiple of 2π. For the whole ring to be superconducting the following condition must be met,

$$\alpha + \beta + 2\pi(\Phi_a/\Phi_0) = n2\pi$$

where α and β are the phase changes produced by currents across the weak-links and $2\pi\Phi_a/\Phi_0$ is the phase change due to the applied magnetic field.

When the measuring current is applied α and β are no longer equal, although their sum must remain constant. The phase changes can be written as

$$\alpha = \pi[n-(\Phi_a/\Phi_0) - \delta$$

$$\beta = \pi[n-(\Phi a/\Phi 0) + \delta$$

where δ is related to the measuring current I. Using the relation between current and phase from the above equation and rearranging to eliminate i we obtain an expression for I,

$$I_c = 2i_c \left| \cos\pi(\Phi_a/\Phi_0).\sin\delta \right|$$

As $\sin\delta$ cannot be greater than unity we can obtain the critical measuring current, Ic from the above,

$$I_c = 2i_c \left| \cos\pi(\Phi_a/\Phi_0) \right|$$

which gives a periodic dependence on the magnitude of the magnetic field, with a maximum when this field is an integer number of fluxons and a minimum at half integer values as shown in Figure 12.23.

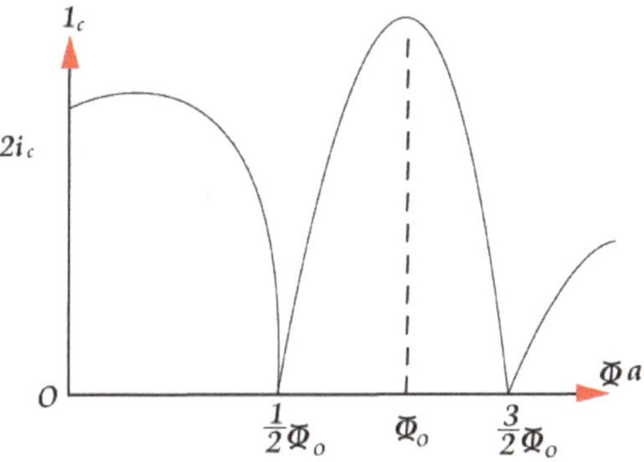

Figure 12.23: Critical measuring current, Ic, as a function of applied magnetic field. Adapted from J. Bland, Thesis M. Phys (Hons), *A Mossbauer Spectroscopy and Magnetometry Study of Magnetic Multilayers and Oxides.* **Oliver Lodge Labs, Department of Physics, University of Liverpool (2002).**

Practical guide to using a superconducting quantum interference device

SQUIDs offer the ability to measure at sensitivities unachievable by other magnetic sensing methodologies. However, their sensitivity requires proper attention to cryogenics and environmental noise. SQUIDs should only be used when no other sensor is adequate for the task. There are many exotic uses for SQUID, however we are just concerned with the laboratory applications.

In most physical and chemical laboratories, a device called a MPMS (Figure 12.24) is used to measure the magnetic moment of a sample by reading the output of the SQUID detector. In a MPMS the sample moves upward through the electronic pick up coils called gradiometers. One upward movement is one whole scan. Multiple scans are used and added together to improve measurement resolution. After collecting the raw voltages, there is computation of the magnetic moments of the sample.

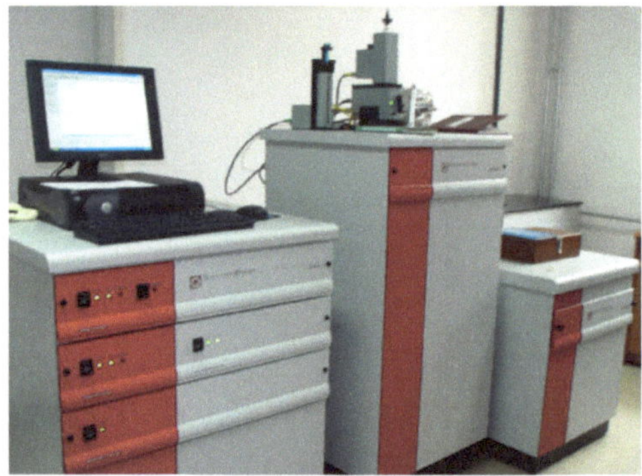

Figure 12.24: A MPMS workstation.

The MPMS measures the moment of a sample by moving it through a liquid Helium cooled, superconducting sensing coil. Many different measurements can be carried out using an MPMS however we will discuss just a few.

Using a magnetic property measurement system (MPMS)

DC magnetization

DC magnetization is the magnetic per unit volume (M) of a sample. If the sample doesn't have a permanent magnetic moment, a field is applied to induce one. The sample is then stepped through a superconducting detection array and the SQUID's output voltage is processed and the sample moment computed. Systems can be configured to measure hysteresis loops, relaxation times, magnetic field, and temperature dependence of the magnetic moment.

A DC field can be used to magnetize samples. Typically, the field is fixed, and the sample is moved into the detection coil's region of sensitivity. The change in detected magnetization is directly proportional to the magnetic moment of the sample. Commonly referred to as SQUID magnetometers, these systems are properly called SQUID susceptometers (Figure 12.25). They have a homogeneous superconducting magnet to create a very uniform field over the entire sample measuring region and the superconducting pickup loops. The magnet induces a moment allowing a measurement of magnetic susceptibility. The superconducting detection loop array is rigidly mounted in the center of the magnet. This array is configured as a gradient coil to reject external noise sources. The detection coil geometry determines what mathematical algorithm is used to calculate the net magnetization.

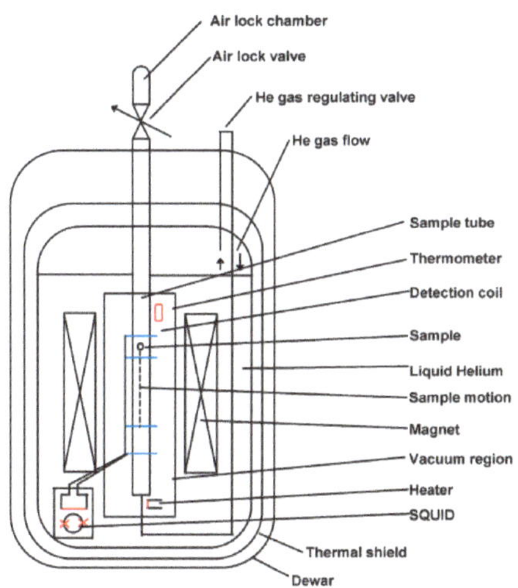

Figure 12.25: Schematic diagram of a MPMSR. Adapted from L. Fagaly, Superconducting quantum interference device instruments and applications. *Rev. Sci.*, 2006, 77, 101101. Copyright: American Institute of Physics (2006).

An important feature of SQUIDs is that the induced current is independent of the rate of flux change. This provides uniform response at all frequencies i.e., true dc response and allows the sample to be moved slowly without degrading performance. As the sample passes through a coil, it changes the flux in that coil by an amount proportional to the magnetic moment M of the sample. The peak-to-peak signal from a complete cycle is thus proportional to twice M. The SQUID sensor shielded inside a niobium can is located where the fringe fields generated by the magnet are less than 10 mT. The detection coil circuitry is typically constructed using NbTi (Figure 12.25). This allows measurements in applied fields of 9 T while maintaining sensitivities of $10-8$ emu. Thermal insulation not shown is placed between the detection coils and the sample tube to allow the sample temperature to be varied.

The use of a variable temperature insert can allow measurements to be made over a wide range 1.8 - 400 K. Typically, the sample temperature is controlled by helium gas flowing slowly past the sample. The temperature of this gas is regulated using a heater located below the sample measuring region and a thermometer located above the sample region. This arrangement ensures that the entire region has reached thermal equilibrium prior to data acquisition. The helium gas is obtained from normal evaporation in the Dewar, and its flow rate is controlled by a precision regulating valve.

Procedures when using an MPMS

Calibration

The magnetic moment calibration for the SQUID is determined by measuring a palladium standard over a range of magnetic fields and then by adjusting to obtain the correct moment for the standard. The palladium standard samples are effectively point sources with an accuracy of approximately 0.1%.

Sample mounting considerations

The type, size and geometry of a sample is usually sufficient to determine the method you use to attach it to the sample. However mostly for MPMS measurements a plastic straw is used. This is due to the straw having minimal magnetic susceptibility.

However, there are a few important considerations for the sample holder design when mounting a sample for measurement in a magnetometer. The sample holder can be a major contributor to the background signal. Its contribution can be minimized by choosing materials with low magnetic susceptibility and by keeping the mass to a minimum such as a plastic straw mentioned above.

The materials used to hold a sample must perform well over the temperature range to be used. In a MPMS, the geometric arrangement of the background and sample is critical when their magnetic susceptibilities will be of similar magnitude. Thus, the sample holder should optimize the sample's positioning in the magnetometer. A sample should be mounted rigidly in order to avoid excess sample motion during measurement. A sample holder should also allow easy access for mounting the sample, and its background contribution should be easy to measure. This advisory introduces some mounting methods and discusses some of the more important considerations when mounting samples for the MPMS magnetometer. Keep in mind that these are only recommendations, not guaranteed procedures. The researcher is responsible for assuring that the methods and materials used will meet experimental requirements.

Sample Mounts

Platform mounting: For many types of samples, mounting to a platform is the most convenient method. The platform's mass and susceptibility should be as small as possible in order to minimize its background contribution and signal distortion.

Plastic disc: A plastic disc about 2 mm thick with an outside diameter equivalent to the pliable plastic tube's diameter (a clear drinking straw is suitable) is inserted and twisted into place. The platform should be fairly rigid. Mount samples onto this platform with glue. Place a second disc, with a diameter slightly less than the inside diameter of the tube and with the same mass, on top of the sample to help provide the desired symmetry. Pour powdered samples onto the platform and place a second disc on top. The powders will be able to align with the field. Make sure the sample tube is capped and ventilated.

Crossed threads: Make one of the lowest mass sample platforms by threading a cross of white cotton thread (colored dyes can be magnetic). Using a needle made of a non-magnetic metal, or at least carefully cleaned, thread some white cotton sewing thread through the tube walls and tie a secure knot so that the thread platform is rigid. Glue a sample to this platform or use the platform as a support for a sample in a container. Use an additional thread cross on top to hold the container in place.

Gelatin capsule: Gelatin capsules can be very useful for containing and mounting samples. Many aspects of using gelatin capsules have been mentioned in the section, Containing the Sample. It is best if the sample is mounted near the capsule's center, or if it completely fills the capsule. Use extra capsule parts to produce mirror symmetry. The thread cross is an excellent way of holding a capsule in place.

Thread mounting: Another method of sample mounting is attaching the sample to a thread that runs through the sample tube. The thread can be attached to the sample holder at the ends of the sample tube with tape, for example. This method can be very useful with at samples, such as those on substrates, particularly when the field is in the plane of the film. Be sure to close the sample tube with caps.
- Mounting with a disc platform.
- Mounting on crossed threads.
- Long thread mounting.

Steps for inserting the sample

1. Cut off a small section of a clear plastic drinking straw. The section must be small enough to fit inside the straw.
2. Weigh and measure the sample.
3. Use plastic tweezers to place the sample inside the small straw segment. It is important to use plastic tweezers not metallic ones as these will contaminate the sample.
4. Place the small straw segment inside the larger one. It should be approximately in the middle of the large drinking straw.
5. Attach the straw to the sample rod which is used to insert the sample into the SQUID machine.
6. Insert the sample rod with the attached straw into the vertical insertion hole on top of the SQUID.

Center the sample

The sample must be centered in the SQUID pickup coils to ensure that all coils sense the magnetic moment of the sample. If the sample is not centered, the coils read only part of the magnetic moment.

During a centering measurement the MPMS scans the entire length of the samples vertical travel path, and the MPMS reads the maximum number of data points. During centering there are a number of terms which need to be understood.

- A scan length is the length of a scan of a particular sample which should usually try and be the maximum of the sample.
- A sample is centered when it is in the middle of a scan length. The data points are individual voltage readings plotting response curves in centering scan data les.
- Auto-tracking is the adjustment of a sample position to keep a sample centered in SQUID coils. Auto- tracking compensates for thermal expansion and contraction in a sample rod.

As soon as a centering measurement is initiated, the sample transport moves upward, carrying the sample through the pickup coils. While the sample moves through the coils, the MPMS measures the SQUID's response to the magnetic moment of the sample and saves all the data from the centering measurement.

After a centering plot is performed the plot is examined to determine whether the sample is centered in the SQUID pickup coils. The sample is centered when the part of the large, middle curve is within 5cm of the half-way point of the scan length.

The shape of the plot is a function of the geometry of the coils. The coils are wound in a way which strongly rejects interference from nearby magnetic sources and lets the MPMS function without a superconducting shield around the pickup coils.

Geometric considerations

To minimize background noise and stray field effects, the MPMS magnetometer pick-up coil takes the form of a second-order gradiometer. An important feature of this gradiometer is that moving a long, homogeneous sample through it produces no signal as long as the sample extends well beyond the ends of the coil during measurement.

As a sample holder is moved through the gradiometer pickup coil, changes in thickness, mass, density, or magnetic susceptibility produce a signal. Ideally, only the sample to be measured produces this change. A homogeneous sample that extends well beyond the pick-up coils does not produce a signal, yet a small sample does produce a signal. There must be a crossover between these two limits. The sample length (along the field direction) should not exceed 10 mm. In order to obtain the most accurate measurements, it is important to keep the sample susceptibility constant over its length; otherwise distortions in the SQUID signal (deviations from a dipole signal) can result. It is also important to keep the sample close to the magnetometer centerline to get the most accurate measurements. When the sample holder background contribution is similar in magnitude to the sample signal, the relative positions of the sample and the materials producing the background are important. If there is a spatial offset between the two along the magnet axis, the signal produced by the combined sample and background can be highly distorted and will not be characteristic of the dipole moment being measured.

Even if the signal looks good at one temperature, a problem can occur if either of the contributions are temperature dependent. Careful sample positioning and a sample holder with a center, or plane, of symmetry at the sample (i.e., materials distributed symmetrically about the sample, or along the principal axis for a symmetry plane) helps eliminate problems associated with spatial offsets.

Containing the Sample

Keep the sample space of the MPMS magnetometer clean and free of contamination with foreign materials. Avoid accidental sample loss into the sample space by properly containing the sample in an appropriate sample holder. In all cases it is important to close the sample holder tube with caps in order to contain a sample that might become unmounted. This helps avoid sample loss and subsequent damage during the otherwise unnecessary recovery procedure. Position caps well out of the sample-measuring region and introduce proper venting.

Sample preparation workspace

Work area cleanliness and avoiding sample contamination are very important concerns. There are many possible sources of contamination in a laboratory. Use diamond tools when cutting hard materials. Avoid carbide tools because of potential contamination by the cobalt binder found in many carbide materials. The best tools for preparing samples and sample holders are made of plastic, titanium, brass, and beryllium copper (which also has a small amount of cobalt). Tools labeled non-magnetic can actually be made of steel and often be made "magnetic" from exposure to magnetic fields. However, the main concern from these "non-magnetic" tools is contamination by the iron and other ferrous metals in the tool. It is important to have a clean white-papered workspace and a set of tools dedicated to mounting your own samples. In many cases, the materials and tools used can be washed in dilute acid to remove ferrous metal impurities. Follow any acid washes with careful rinsing with deionized water.

Powdered samples pose a special contamination threat, and special precautions must be taken to contain them. If the sample is highly magnetic, it is often advantageous to embed it in a low susceptibility epoxy matrix, or a nitrocellulose-based cement (Figure 12.26) such as Duco®. This is usually done by mixing a small amount of diluted glue with the powder in a suitable container such as a gelatin capsule. Potting the sample in this way can keep the sample from shifting or aligning with the magnetic field. In the case of weaker magnetic samples, measure the mass of the glue after drying and making a background measurement. If the powdered sample is not potted, seal it into a container, and watch it carefully as it is cycled in the airlock chamber.

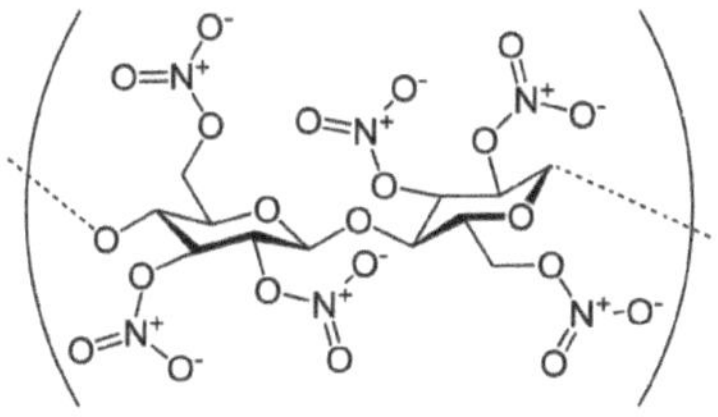

Figure 12.26: Structure of the nitrocellulose component of Duco Cement®.

Pressure equalization

The sample space of the MPMS has a helium atmosphere maintained at low pressure of a few torr. An airlock chamber is provided to avoid contamination of the sample space with air when introducing samples into the sample space. By pushing the purge button, the airlock is cycled between vacuum and helium gas three times, then pumped down to its working pressure. During the cycling, it is possible for samples to be displaced in their holders, sealed capsules to explode, and sample holders to be deformed. Many of these problems can be avoided if the sample holder is properly ventilated. This requires placing holes in the sample holder, out of the measuring region that will allow any closed spaces to be opened to the interlock chamber.

Air-sensitive samples and liquid samples

When working with highly air-sensitive samples or liquid samples it is best to first seal the sample into a glass tube. NMR and EPR tubes make good sample holders since they are usually made of a high-quality, low-susceptibility glass or fused silica. When the sample has a high susceptibility, the tube with the sample can be placed onto a platform like those described earlier. When dealing with a low susceptibility sample, it is useful to rest the bottom of the sample tube on a length of the same type of glass tubing. By producing near mirror symmetry, this method gives a nearly constant background with position and provides an easy method for background measurement (i.e., measure the empty tube first, then measure with a sample). Be sure that the tube ends are well out of the measuring region.

When going to low temperatures, check to make sure that the sample tube will not break due to differential thermal expansion. Samples that will go above room temperature should be sealed with a reduced pressure in the tube and be checked by taking the sample to the maximum experimental temperature prior to loading it into the magnetometer. These checks are especially important when the sample may be corrosive, reactive, or valuable.

Oxygen contamination

This application note describes potential sources for oxygen contamination in the sample chamber and discusses its possible effects. Molecular oxygen, which undergoes an antiferromagnetic transition at about 43 K, is strongly paramagnetic above this temperature. The MPMS system can easily detect the presence of a small amount of condensed oxygen on the sample, which when in the sample chamber can interfere significantly with sensitive magnetic measurements. Oxygen contamination in the sample chamber is usually the result of leaks in the system due to faulty seals, improper operation of the airlock valve, outgassing from the sample, or cold samples being loaded.

Bibliography

G. Bertotti, *Hysteresis in Magnetism*, Academic Press, San Diego (1998).

J. Bland, Thesis M. Phys (Hons), *A Mossbauer Spectroscopy and Magnetometry Study of Magnetic Multilayers and Oxides*. Oliver Lodge Labs, Department of Physics, University of Liverpool (2002).

S. Chikazumi and S. H. Charap, *Physics of Magnetisim*, Krieger Publishing Company (1978).

J. P. Cleuziou, W. Wernsdorfer, V. Bouchiat, T. Ondarçuhu, and M. Monthioux, Carbon nanotube superconducting quantum interference device. *Nat, Nanotechol.*, 2006, **1**, 53.

B. D. Cullity, *Introduction to Magnetic Materials*, Addison-Wesley, Massachusettes (1972).

R. L. Fagaly, Superconducting quantum interference device instruments and applications. *Rev. Sci.*, 2006, **77**, 101101.

W. G. Jenks, S. S. H. Sadeghi, and J. P. Wikswo, Jr., SQUIDs for nondestructive evaluation. *J. Phys. D: Appl. Phys.*, 1997, **30**, 293.

L. Morrow, D. Potter, and A. R. Barron, Detection of magnetic nanoparticles against proppant and shale reservoir rocks. *J. Exp. Nanosci.*, 2014, **9**, 1028

L. Morrow, B. Snow, A. Ali, S. J. Maguire-Boyle, Z. Almutairi, D. K. Potter, and A. R. Barron, Temperature dependence on the mass susceptibility and mass magnetization of superparamagnetic Mn-Zn-ferrite nanoparticles as contrast agents for magnetic imaging of oil and gas reservoirs. *J. Exp. Nanosci.*, 2018, **13**, 107

Quantum Design, *Operating Manual for the MPMS*, 1999.

U. Gradmann, in *Handbook of Magnetic Materials*, Vol. 7. K. H. J. Buschow, Ed. North Holland, Amsterdam (1993).

234

Chapter 13: Low-Temperature Specific Heat Measurements for Magnetic Materials

Kyle Bayliff, Pavan M. V. Raja and Andrew R. Barron

Introduction

Magnetic materials attract the attention of researchers and engineers because of their potential for application in magnetic and electronic devices such as navigational equipment, computers, and even high-speed transportation. Perhaps more valuable still, however, is the insight they provide into fundamental physicals. Magnetic materials provide an opportunity for studying exotic quantum mechanical phenomena such as quantum criticality, superconductivity, and heavy fermionic behavior intrinsic to these materials. A battery of characterization techniques exists for measuring the physical properties of these materials, among them a method for measuring the specific heat of a material throughout a large range of temperatures. Specific heat measurements are an important means of determining the transition temperature of magnetic materials the temperature below which magnetic ordering occurs. Additionally, the functionality of specific heat with temperature is characteristic of the behavior of electrons within the material and can be used to classify materials into different categories.

Temperature-dependence of specific heat

The molar specific heat of a material is defined as the amount of energy required to raise the temperature of 1 mole of the material by 1 K. This value is calculated theoretically by taking the partial derivative of the internal energy with respect to temperature. This value is not a constant, as it is typically treated in high-school science courses: it depends on the temperature of the material. Moreover, the temperature- dependence itself also changes based on the type of material. There are three broad families of solid-state materials defined by their specific heat behaviors. Each of these families is discussed in the following sections.

Insulators

Insulators have specific heat with the simplest dependence on temperature. According to the Debye (Figure 13.1) theory of specific heat, which models materials as phonons (lattice vibrational modes) in a potential well, the internal energy of an insulating system is given by,

$$U = \frac{9Nk_BT^4}{T_D^3} \int_0^{T_D/T} \frac{x^3}{e^x - 1}\, dx$$

where T_D is the Debye temperature, defined as the temperature associated with the energy of the highest allowed phonon mode of the material. In the limit that $T \ll T_D$, the energy expression reduces to,

$$U = \frac{3\pi^4 Nk_BT^4}{5T_D^3}$$

Figure 13.1: Dutch physicist and physical chemist Peter Joseph William Debye (1884 - 1966) recipient of the Nobel Prize in Chemistry.

For most magnetic materials, the Debye temperature is several orders of magnitude higher than the temperature at which magnetic ordering occurs, making this a valid approximation of the internal energy. The specific heat derived from this expression is given by,

$$C_v = \frac{\partial U}{\partial T} = \frac{12\pi^4 Nk_B}{5T_D^3} T^3 = \beta T^3$$

The behavior described by the Debye theory accurately matches experimental measurements of specific heat for insulators at low temperatures. Normal insulators, then, have a T^3 dependence in the specific heat that is dominated by contributions from phonon excitations. Essentially all energy absorbed by insulating materials is stored in the vibrational modes of a solid lattice. At very low temperatures this contribution is very small, and insulators display a high sensitivity to changes in heat energy.

Metals: Fermi liquids

While the Debye theory of specific heat accurately describes the behavior of insulators, it does not adequately describe the temperature dependence of the specific heat for metallic materials at low temperatures, where contributions from delocalized conduction electrons becomes significant. The predictions made by the Debye model are corrected in the Einstein-Debye model of specific heat, where an additional term describing the contributions from the electrons (as modeled by a free electron gas) is added to the phonon contribution. The internal energy of a free electron gas is given by,

$$U = \frac{\pi^2}{6}(k_B T)^2 g\left(E_f\right) + U_0$$

where $g(E_f)$ is the density of states at the Fermi level, which is material dependent. The partial derivative of this expression with respect to temperature yields the specific heat of the electron gas,

$$C_v = \frac{\pi^2}{3} k_B^2 g\left(E_f\right) T = \gamma T$$

Combining this expression with the phonon contribution to specific heat gives the expression predicted by the Einstein-Debye model,

$$C_v = \frac{\pi^2}{3} k_B^2 g\left(E_f\right) T + \frac{12\pi^4 N k_B}{5 T_D^3} T^3 = \gamma T + \beta T^3$$

This is the general expression for the specific heat of a Fermi liquid a variation on the Fermi gas in which fermions (typically electrons) are allowed to interact with each other and form quasiparticles weakly bound and often short-lived composites of more fundamental particles such as electron-hole pairs or the Cooper pairs of BCS superconductor theory.

Most metallic materials follow this behavior and are thus classified as Fermi liquids. This is easily confirmed by measuring the heat capacity as a function of temperature and linearizing the results by plotting C/T versus T^2. The slope of this graph equals the coefficient β, and the y-intercept is equal to γ. The ability to obtain these coefficients is important for gaining understanding of some unique physical phenomena. For example, the compound $YbRh_2Si_2$ (Figure 13.2) is a heavy fermionic material a material with charge carriers that have an effective mass much greater than the normal mass of an electron. The increased mass is due to coupling of magnetic moments between conduction electrons and localized magnetic ions. The coefficient γ is related to the density of states at the Fermi level, which is dependent on the carrier mass. Determination of this coefficient via specific heat measurements provides a way to determine the effective carrier mass and the coupling strength of the quasiparticles.

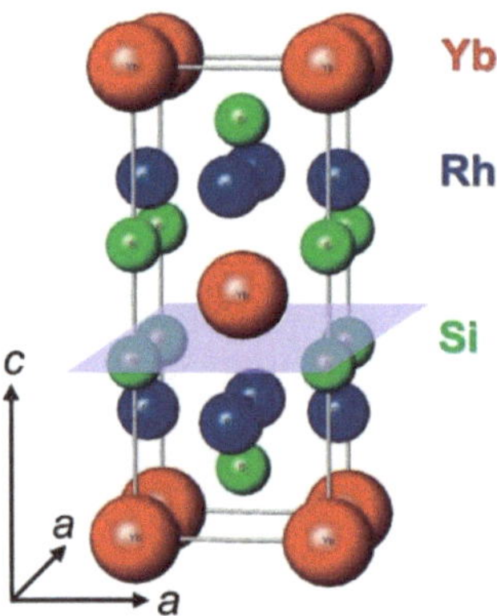

Figure 13.2: Structure of heavy fermion metal, tetragonal $YbRh_2Si_2$.

Additionally, knowledge of Fermi-liquid behavior provides insight for application development. The temperature dependence of the specific heat shows that the phonon contribution dominates at higher temperatures, where the behavior of metals and insulators is very similar. At low temperatures, the electronic term is dominant, and metals can absorb more heat without a significant change in temperature. As will be discussed briefly later, this property of metals is utilized in low-temperature refrigeration systems for heat storage at low temperatures.

Metals: non-Fermi liquids

While most metals fall under the category of Fermi liquids, there are some that show a different dependence on temperature. Naturally, these are classified as non-Fermi liquids. Often, deviation from Fermi-liquid behavior is an indicator of some of the interesting physical phenomena that currently garner the attention of many condensed matter researchers. For instance, non-Fermi liquid behavior has been observed near quantum critical points. Classically, fluctuations in physical properties such as magnetic susceptibility and resistivity occur near critical points which include phase changes or magnetic ordering transitions. Normally, these fluctuations are suppressed at low temperatures at absolute zero, classical systems collapse into the lowest energy state and remain stable; However, when the critical transition temperature is lowered by the application of pressure, doping, or magnetic field to absolute zero, the fluctuations are enhanced as the temperature approaches absolute zero, propagating throughout the whole of the material. As this is not classically allowed, this behavior indicates a quantum mechanical effect at play that is currently not well understood. The transition point is then called a quantum critical point. Non-fermi liquid behavior as identified by deviations in the expected specific heat, then, is used to identify materials that can provide an experimental basis for development of a theory that describes the physics of quantum criticality.

Determination of magnetic transition temperatures via specific heat measurements

While analysis of the temperature dependence of specific heat is a vital tool for studying the strange physical behaviors of quantum mechanics in solid state materials, these are studied by only a small subsection of the physics community. The utility of specific heat measurements is not limited to a few niche subjects, however. Possibly the most important use for specific heat measurements is the determination of critical transition temperatures. For any sort of physical state transition phase transitions, magnetic ordering, transitions to superconducting states a sharp increase in the specific heat occurs during the transition. This increase in specific heat is the reason why, for example, water does not change temperature as it changes from a liquid to a solid. These increases are quite obvious in plots of the specific heat versus temperature as seen in Figure 13.3. These transition-associated peaks are called Schottky anomalies, as normal specific heat behavior is not followed near to the transition temperature.

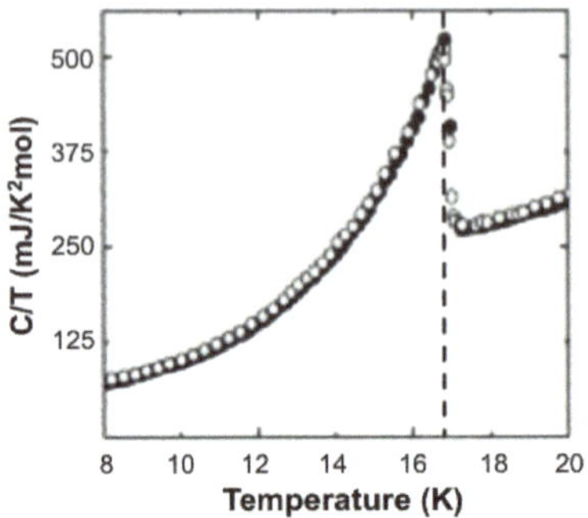

Figure 13.3: A Schottky anomaly at the magnetic ordering temperature (T_c = 16.8 K) of URu$_2$Si$_2$ (Figure 13.4). Adapted from J. C. Lashley, M. F. Hundley, A. Migliori, J. L. Sarrao, P. G. Pagliuso, T. W. Darling, M. Jaime, J. C. Cooley, W. L. Hults, L. Morales, D. J. Thoma, J. L. Smith, J. Boerio-Goates, B. F. Wood field, G. R.Stewart, R. A. Fisher, and N. E. Phillips, Critical examination of heat capacity measurements made on a quantum design physical property measurement system. *Cryogenics*, 2003, 43, 369. Copyright: Elsevier (2003).

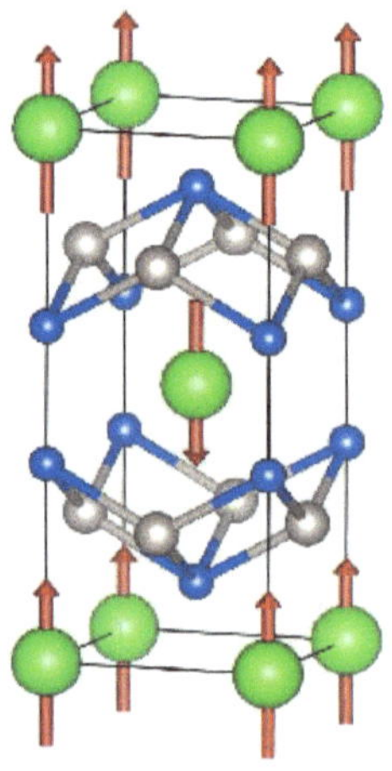

Figure 13.4: The unit cell structure of URu$_2$Si$_2$. Atom colors: U = green; Ru = silver; Si – blue. Adapted from Y. L. Wang, G. Fabbris, D. Meyers, N. H. Sung, R. E. Baumbach, E. D. Bauer, P. J. Ryan, J.-W. Kim, X. Liu, M. P. M. Dean, G. Kotliar, and X. Dai, On the possibility to detect multipolar order in URu$_2$Si$_2$ by the electric quadrupolar transition of resonant elastic X-ray scattering. Phys. Rev. B, 2017, 96, 085146. Copyright: American Physical Society (2017).

For the purposes of this chapter, the following sections will focus on specific heat measurements as they relate to magnetic ordering transitions. The following sections will describe the practical aspects of measuring the specific heat of these materials.

A practical guide to low-temperature specific heat measurements

The thermal relaxation method of measurement

Specific heat is measured using a calorimeter. The design of basic calorimeters for use over a short range of temperatures is relatively simple. They consist of a sample with a known mass and an unknown specific heat, an energy source which provides heat energy to the sample, a heat reservoir (of known mass and specific heat) that absorbs heat from the sample, insulation to provide adiabatic conditions inside the calorimeter, and probes for measuring the temperature of the sample and the reservoir. The sample is heated with a pulse to a temperature higher than the heat reservoir, which decreases as energy is absorbed by the reservoir until a thermal equilibrium is established. The total energy change is calculated using the specific heat and temperature change of the reservoir. The specific heat of the sample is calculated by dividing the total energy change by the product of the mass of the sample and the temperature change of the sample.

However, this method of measurement produces an average value of the specific heat over the range of the change in temperature of the sample, and therefore, is insufficient for producing accurate measurements of the specific heat as a function of temperature. The solution, then, is to minimize the temperature change by reducing the amount of heat added to the system; yet, this presents another obstacle to making measurement as, in general, the temperature change of the reservoir is much smaller than that of the sample. If the change in temperature of the sample is minimized, the temperature change of reservoir becomes too small to measure with precision. A more direct method of measurement, then, seems to be required.

Fortunately, such a method exists: it is known as the thermal relaxation method. This method involves measurement of the specific heat without the need for precise knowledge of temperature changes in the reservoir. In this method, solid samples are a fixed to a platform. Both the specific heat of the sample and the platform itself contribute to the measured specific heat;

therefore, the contribution from the platform must be subtracted. This contribution is determined by measuring the specific heat without a sample present. Both the sample and the platform are in thermal contact with a heat reservoir at low temperature as depicted in Figure 13.5.

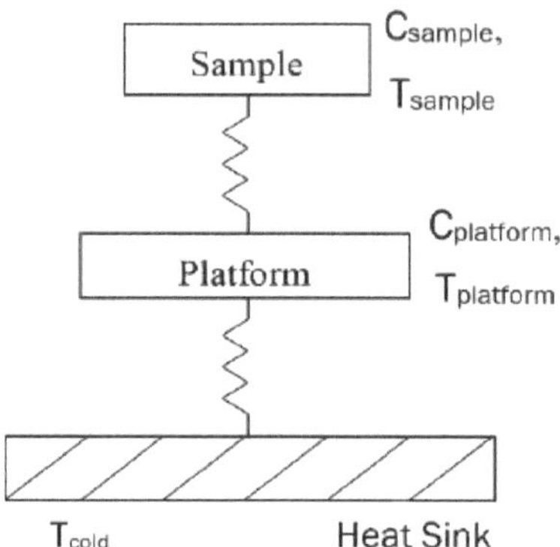

Figure 13.5: A schematic representation depicting the thermal connection between the sample and the heat reservoir.

A heat pulse is delivered to the sample to produce a minimal increase in the temperature of the sample. The temperature is measured versus time as it decays back to the temperature of the reservoir as shown in Figure 13.6.

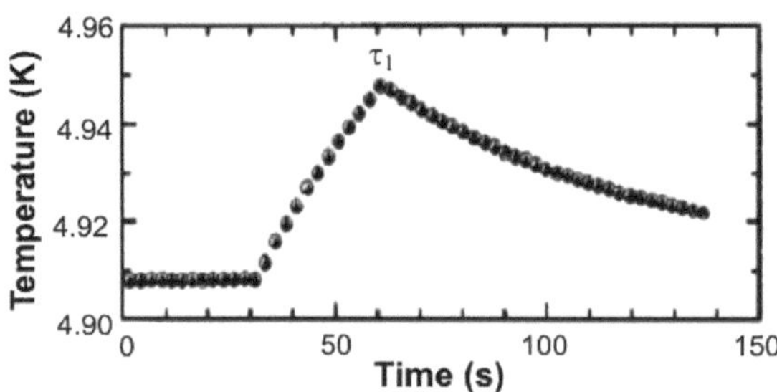

Figure 13.6: An example of a low-temperature heat-pulse temperature decay for a copper standard, τ_1 = 65.5 s. Adapted from J. S. Hwang, K. J. Lin, and C. Tien, Measurement of heat capacity by fitting the whole temperature response of a heat-pulse calorimeter. *Rev. Sci.*, 1997, 68, 94. Copyright: American Institute of Physics (1997).

The temperature of the sample decays according to,

$$T = \Delta T e^{-t/\tau} + T_0$$

where T_0 is the temperature of the heat reservoir, and ΔT is the temperature difference between the initial sample temperature and the reservoir temperature. The decay time constant τ is directly related to the specific heat of the sample by,

$$\tau = C_p/K$$

where K is the thermal conductance of the thermal link between the sample and the heat reservoir. In order for this to be valid, however, the thermal conductance must be sufficiently large that the energy transfer from the heated sample to the reservoir can be treated as a single process. If the thermal conduction is poor, a two-τ behavior arises corresponding to two separate processes with different time constants slow heat transfer from the sample to the platform, and fast transfer from the platform to the reservoir. Figure 13.7 shows a relaxation curve in which the two-τ behavior plays a significant role.

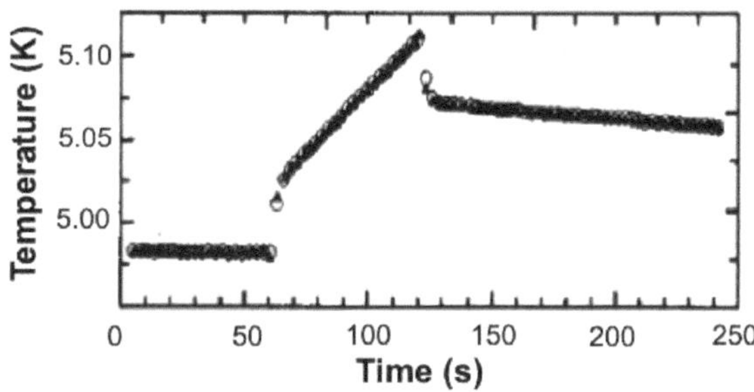

Figure 13.7: A thermal relaxation decay graph for a sample of a Pb sample displaying the two-τ effects. J. S. Hwang, K. J. Lin, and C. Tien, Measurement of heat capacity by fitting the whole temperature response of a heat-pulse calorimeter. *Rev. Sci.*, 1997, 68, 94. Copyright: American Institute of Physics (1997).

The two-τ effect is generally undesirable for making measurements. It can be avoided by reducing thermal conductance between the sample and the platform, effectively making the contribution from the heat transfer from the sample to the platform insignificant compared to the transfer from the platform to the reservoir; however, if the conductance between the sample and the platform is too low, the time required to reach thermal equilibrium becomes excessively long, translating into very long measurement times. It is necessary, then, to optimize the conductance to compensate for both of these issues. This essentially provides a limitation on the temperature range over which these effects are insignificant.

In order to measure at different temperatures, the temperature of the heat reservoir is increased stepwise from the lowest temperature until the desired temperature range is covered. At each step, the temperature is allowed to equilibrate, and a data point is measured.

Instrumentation

Thermal relaxation calorimeters use advanced technology to make precise measurements of the specific heat using components made of highly specialized materials. For example, the sample platform is made of synthetic sapphire which is used as a standard material, the grease which is applied to the sample to provide even thermal contact with the platform is a special hydrocarbon-based material which can withstand millikelvin temperatures without creeping, cracking, or releasing vapor, and the resistance thermometers used for ultralow temperatures are often made of treated graphite or germanium. The culmination of years of materials science research and careful engineering has produced instrumentation with the capability for precise measurements from temperatures down to the millikelvin level. There are four main systems that function to provide the proper conditions for measurement: the reservoir temperature control, the sample temperature control, the magnetic field control, and the pressure control system. The essential components of these systems will be discussed in more detail in the following sections with special emphasis on the cooling systems that allow these extreme low temperatures to be achieved.

Cooling systems

The first of these is responsible for maintaining the low baseline temperature to which the sample temperature relaxes. This is typically accomplished with the use of liquid helium cryostats or, in more recent years, so-called cryogen-free pulse tube coolers.

A cryostat is simply a bath of cryogenic fluid that is kept in thermal contact with the sample. The fluid bath may be static or may be pumped through a circulation system for better cooling. The cryostat must also be thermally insulated from the external environment in order to maintain low temperatures. Insulation is provided by a metallic vacuum Dewar. The vacuum virtually eliminates conductive or convective heat transfer from the environment and the reflective metallic outer sleeve acts as a radiation shield. For the low temperatures required to observe some magnetic transitions, liquid helium is generally required. ^{4}He liquefies at 4.2 K, and the rarer (and much more expensive) isotope, ^{3}He, liquefies at 1.8 K. For temperatures lower than 1.8 K, modern instruments employ evaporative attachments such as a 1-K pot, ^{3}He refrigerator, or a dilution refrigerator. The 1-K pot is so named because it can achieve temperatures down to 1 K. It consists of a small vessel filled with liquid ^{4}He under reduced pressure. Heat is absorbed as the liquid evaporates and is carried away by the vapor. The ^{3}He refrigerator utilizes a 1-K pot for liquefaction of ^{3}He, then evaporation of ^{3}He provide cooling to the sample. ^{3}He refrigerators can provide temperatures as low as 200 mK. The dilution refrigerator works on a similar principle; however, the working fluid is a mixture of ^{3}He and ^{4}He. Phase separation of the ^{3}He from the mixture provides further heat absorption as the ^{3}He evaporates. Dilution refrigerators can achieve temperatures as low as 0.002 K. Evaporative refrigerators work only on a small area in thermal contact with the sample, rather than delivering cooling power to the entire volume of the cryostat bath.

Cryostat baths provide very high cooling power for very efficient cooling; however, they come with a major drawback: the cost of helium is prohibitively high. The helium vapor that boils off as it provides cooling to the sample must leave the system in order to carry the heat away and must therefore be replaced. Even when the instrument is not in use, there is some loss of helium due to the imperfect nature of the insulating Dewar. In order to get the best use out of the helium cryostat systems must always be in use. In addition, rather than allowing expensive helium to simply escape, recovery systems for helium exhaust must be installed in order to operate in a cost-effective manner, though these systems are not 100% efficient, and the cost of operation and maintenance of recovery systems is not small either. Cryogen-free coolers provide an alternative to cryostats in order to avoid the costs associated with helium usage and recovery.

Figure 11.8 shows a Gifford-McMahon type pulse tube one example of the cryogen-free coolers. In this type of cooler, helium gas is driven through the regenerator by a compressor. As a small volume element of the gas passes throughout the regenerator, it drops in temperature as it deposits heat into the regenerator. The regenerator must have a high specific heat in order to effectively absorb energy from the helium gas. For higher-temperature pulse tube coolers, the regenerator is often made of copper mesh; however, for very low temperatures, helium has a higher specific heat than most metals. Refrigerators for this temperature range are often made of porous rare earth ceramics with magnetic transitions in the low temperature range. The increase in specific heat near the Schottky anomaly for these materials provides the necessary capacity for heat absorption. As the gas enters the tube at a temperature T_L (from Figure 13.8) it is compressed, raising the temperature in accordance with the ideal gas law. At this point, the gas is at a temperature higher than T_H and excess heat is exhausted through the heat exchanger marked X_3 until the temperature is in equilibrium with T_H. When the rotary valve in the compressor turns, the expansion cycle begins, and the gas cools as it expands adiabatically to a temperature below T_L. It then absorbs heat from the sample through the heat exchanger X_2. This step provides the cooling power in pulse tube coolers. Afterward, it travels back through the regenerator at a cold temperature and reabsorbs the heat that was initially stored during compression and regains its original temperature through the heat exchanger X_1. Figure 13.9 illustrates the temperature cycle experienced by a volume element of the working gas as it moves through the pulse tube.

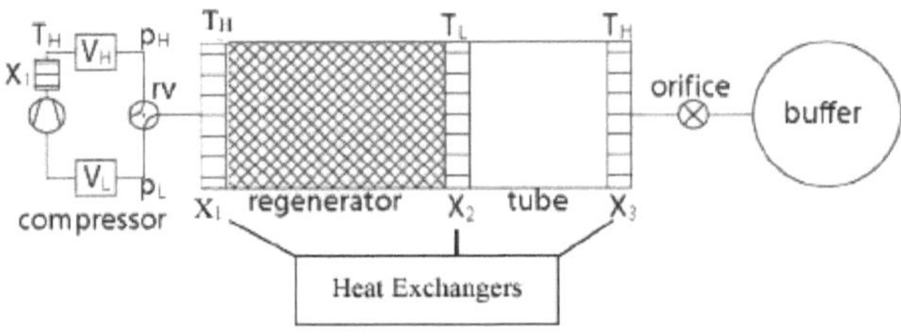

Figure 13.8: A schematic representation of a Gifford-McMahon type pulse tube cooler. Adapted from P. D. Nissen. Closed Cycle Refrigerator Pulse Tube Coolers.

Pulse tube coolers are not truly cryogen-free as they are advertised, but they are preferable to cryostats because there is no net loss of the helium in the system. However, pulse tubes are not a perfect solution. They have very low efficiency over large changes in temperature and at very low temperatures as given by,

$$\xi = 1 - (\Delta T/T_H)$$

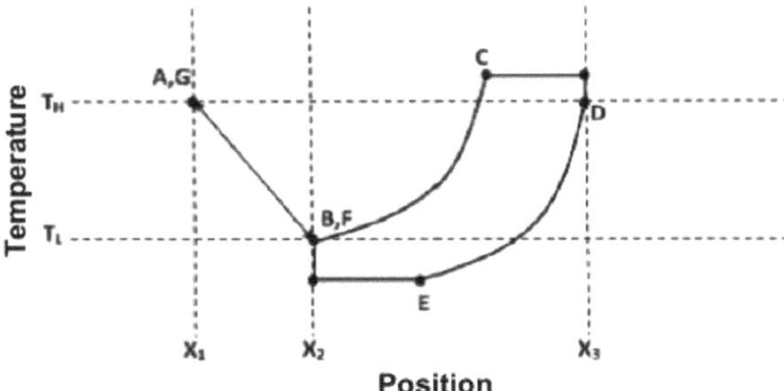

Figure 13.9: Temperature cycle of a gas element moving in the pulse tube cooler. A to B: Gas initially at T_H moves through the regenerator to heat exchanger X_2 dropping to T_L. B to C: Gas is compressed, and the temperature rises above T_H. C to D: Gas is shunted along to X_3 and drops to T_H. D to E: Gas is expanded adiabatically to $T<T_L$. E to F: Cold gas is shunted to X_2 and rises to T_L, absorbing heat from the sample. F to G: Gas at T_L moves through the regenerator reabsorbing heat until it reaches T_H at heat exchanger X_1.

As a result, pulse tube coolers consume a lot of electricity to provide the necessary cooling and may take a long time to achieve the desired temperature. Over large temperature ranges such as the 4 - 300 K range typically used in specific heat measurements, pulse tubes can be used in stages, with one providing pre- cooling for the next, to increase the cooling power and provide a shorter cooling time, though this tends to increase the energy consumption. The cost of running a pulse tube system is still generally less than that of a cryostat, however, and unlike cryostats, pulse tube systems do not have to be used constantly in order to remain cost-effective.

Sample conditions

While the cooling system works more or less independently, the other systems the sample temperature control, the magnetic field control, and the pressure control systems work together to create the proper conditions for measurement of the sample. The sample temperature control system provides the heat pulse used to increase the temperature of the sample before relaxation occurs. The components of this system are incorporated into the sapphire sample platform as shown in Figure 13.10.

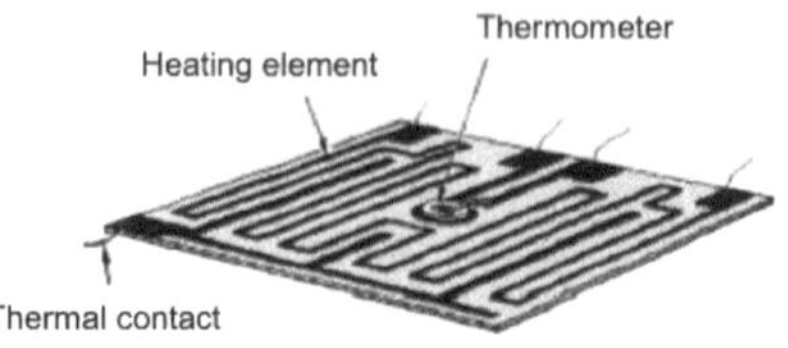

Figure 13.10: The sample platform with important components of the sample temperature control system. Adapted from from R. J. Schutz, Thermal relaxation calorimetry below 1 K. *Rev. Sci.*, 1974, 45, 548. Copyright: American Institute of Physics (1974).

The sample is a fixed to the platform over the thermometer with a small amount of grease, which also provides thermal conductance between the heating element and the sample. The heat pulse is delivered to the sample by running a small current pulse through the heating element, and the response is measured by a resistance thermometer. The resistance thermometer is made of specially treated carbon or germanium which have standardized resistances for given temperatures. The thermometer is calibrated to these standards to provide accurate temperature readings throughout the range of temperatures used for specific heat measurements. A conductive wire provides thermal connection between the sample platform and the heat reservoir. This wire must provide high conductivity to ensure that the heat transfer from the sample to the platform is the dominant process and prevent significant two-τ behavior. Sample preparation is also governed by the temperature control system. The sample must be in good thermal contact with the platform, therefore, a sample with a flat face is preferable. The volume of the sample cannot be too large, either, or the heating element will not be able to heat the sample uniformly. A temperature gradient throughout the sample skews the measurement of the temperature made by the thermometer. Moreover, it is impossible to assign a 1:1 correspondence between the specific heat and temperature if the specific heat values do not correspond to a singular temperature. For the best measurements, heat capacity samples must be cut from large single-crystals or polycrystalline solids using a hard diamond saw to prevent contamination of the sample with foreign material.

The magnetic field control system provides magnetic fields ranging from 0 to >15 T. As was mentioned previously, strong magnetic fields can suppress the transition to magnetically ordered states to lower temperatures, which is important for studying quantum critical behaviors. The magnetic field control consists of a high-current solenoid and regulating electronics to ensure stable current and field outputs.

The pressure systems control the pressure in the sample chamber, which is physically separated from the bath by a wall which allows thermal transfer only. While the sample is installed in the chamber, the vacuum system must be able to maintain low pressures ($\sim 10^{-5}$ Torr) to ensure that no gas is present. If the vacuum system fails, water from any air present in the system can condense inside the sample chamber, including on the sample platform, which alters thermal conductance and throws off measurement of the specific heat. Moreover, as the temperature in the chamber drops, water can freeze and expand in the chamber which can cause significant damage to the instrument itself.

Summary

Through the application of specialized materials and technology, measurements of the specific heat have become both highly accurate and very precise. As our measurement capabilities expand toward the 0 K limit, exciting prospects arise for completion of our understanding, discovery of new phenomena, and development of important applications of novel magnetic materials. Specific heat measurements, then, are a vital tool for studying magnetic materials, whether as a means of exploring the strange phenomena of quantum physics such as quantum criticality or heavy fermions, or simply as a routine method of characterizing physical transitions between different states.

Bibliography

F. J. Blatt. *Modern Physics*, McGraw-Hill, New York (1992).

P. Debye, Zur theorie der spezifischen wärmen. *Ann. Phys.*, 1912, **344**, 789.

A. T. A. M. de Waele, Pulse-tube refrigerators: principle, recent developments, and prospects. *Physica B*, 2000, **280**, 479.

P. Gegenwart, Q. Si, and F. Steglich, Quantum criticality in heavy-fermion metals. *Nature Phys.*, 2006, **4**, 186.

J. S. Hwang, K. J. Lin, and C. Tien, Measurement of heat capacity by fitting the whole temperature response of a heat-pulse calorimeter. *Rev. Sci.*, 1997, **68**, 94.

J. C. Lashley, M. F. Hundley, A. Migliori, J. L. Sarrao, P. G. Pagliuso, T. W. Darling, M. Jaime, J. C. Cooley, W. L. Hults, L. Morales, D. J. Thoma, J. L. Smith, J.

Boerio-Goates, B. F. Wood field, G. R.Stewart, R. A. Fisher, and N. E. Phillips, Critical examination of heat capacity measurements made on a quantum design physical property measurement system. *Cryogenics*, 2003, **43**, 369.

H. V. Löhneysen, A. Rosch, M. Vojta, and P. Wölffe, Fermi-liquid instabilities at magnetic quantum phase transitions. *Rev. Mod. Phys.*, 2007, **79**, 1015.

A. J. Schofield, Non-Fermi liquids. *Contemp. Phys.*, 1999, **40**, 95.

R. J. Schutz, Thermal relaxation calorimetry below 1 K. *Rev. Sci.*, 1974, **45**, 548.

Y. L. Wang, G. Fabbris, D. Meyers, N. H. Sung, R. E. Baumbach, E. D. Bauer, P. J. Ryan, J.-W. Kim, X. Liu, M. P. M. Dean, G. Kotliar, and X. Dai, On the possibility to detect multipolar order in URu_2Si_2 by the electric quadrupolar transition of resonant elastic X-ray scattering. *Phys. Rev. B*, 2017, **96**, 085146.

Chapter 14: Electrical Permittivity of Aqueous Solutions Across the Frequency Range 0.2 - 3 GHz

Stuart J. Corr and Andrew R. Barron

Introduction

Permittivity (in the framework of electromagnetics) is a fundamental material property that describes how a material will affect, and be affected by, a time-varying electromagnetic field. The parameters of permittivity are often treated as a complex function of the applied electromagnetic field as complex numbers allow for the expression of magnitude and phase. The fundamental equation for the complex permittivity of a substance (ε_s) is given by,

$$\varepsilon_s = \varepsilon'(\omega) - i\varepsilon''(\omega)$$

where ε' and ε are the real and imaginary components, respectively, ω is the radial frequency (rad/s) and can be easily converted to frequency (Hertz, Hz) using,

$$\omega = 2\pi f$$

Specifically, the real and imaginary parameters defined within the complex permittivity equation describe how a material will store electromagnetic energy and dissipate that energy as heat. The processes that influence the response of a material to a time-varying electromagnetic field are frequency dependent and are generally classified as either ionic, dipolar, vibrational, or electronic in nature. These processes are highlighted as a function of frequency in Figure 14.1. Ionic processes refer to the general case of a charged ion moving back and forth in response a time-varying electric field, whilst dipolar processes correspond to the 'flipping' and 'twisting' of molecules, which have a permanent electric dipole moment such as that seen with a water molecule in a microwave oven. Examples of vibrational processes include molecular vibrations (e.g., symmetric and asymmetric) and associated vibrational-rotation states that are infrared (IR) active. Electronic processes include optical and ultra-violet (UV) absorption and scattering phenomenon seen across the UV-visible range.

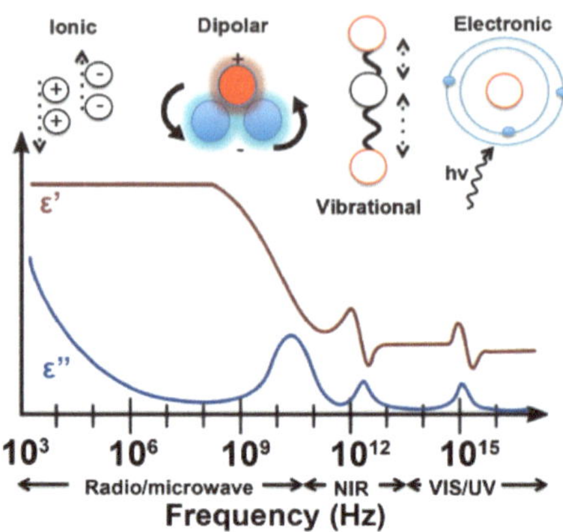

Figure 14.1: A dielectric permittivity spectrum over a wide range of frequencies. ε' and ε" denote the real and the imaginary part of the permittivity, respectively. Various processes are labeled on the image: ionic and dipolar relaxation, and atomic and electronic resonances at higher energies.

The most common relationship scientists that have with permittivity is through the concept of relative permittivity: the permittivity of a material relative to vacuum permittivity. Also known as the dielectric constant, the relative permittivity (ε_r) is given by,

$$\varepsilon_r = \varepsilon_s / \varepsilon_0$$

where ε_s is the permittivity of the substance and ε_0 is the permittivity of a vacuum ($\varepsilon_0 = 8.85 \times 10^{-12}$ Farads/m). Although relative permittivity is dynamic and a function of frequency, the dielectric constants are most often expressed for low frequency electric fields where the electric field is essential static in nature. Table 14.1 depicts the dielectric constants for a range of materials.

Dielectric constants may be useful for generic applications whereby the high-frequency response can be neglected, although applications such as radio communications, microwave design, and optical system design call for a more rigorous and comprehensive analysis. This is especially true for electrical devices such as capacitors, which are circuit elements that store and discharge electrical charge in both a static and time-varying manner. Capacitors can be thought of as two parallel plate electrodes that are separated by a finite distance and 'sandwich' together a piece of material with characteristic permittivity values. As can be seen in Figure 14.2, the capacitance is a function of the permittivity of the material between the plates, which in turn is dependent on frequency. Hence, for capacitors incorporated into the circuit design for radio communication applications, across the spectrum 8.3 kHz 300 GHz, the frequency response would be important as this will determine the capacitors ability to charge and discharge as well as the thermal response from electric fields dissipating their power as heat through the material.

Material	Relative Permittivity
Vacuum	1 (by definition)
Air	1.00058986
Polytetrafluoroethylene (PTFE, Teflon)	2.1
Paper	3.85
Diamond	5.5 - 10
Methanol	30
Water	80.1
Titanium dioxide (TiO_2)	86 - 173
Strontium titanate ($SrTiO_3$)	310
Barium titanate ($BaTiO_3$)	1,200-10,000
Calcium copper titanate ($CaCu_3Ti_4O_{12}$)	> 250,000

Table 14.1: Relative permittivities of various materials under static (i.e., non time-varying) electric fields.

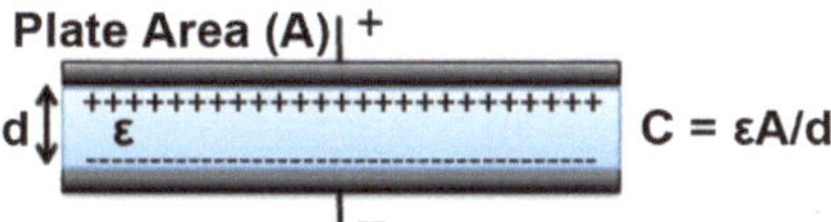

Figure 14.2: Parallel plate capacitor of area, A, separated by a distance, d. The capacitance of the capacitor is directly related to the permittivity (ε) of the material between the plates, as shown in the equation.

Evaluating the electrical characteristics of materials is become increasingly popular especially in the field of electronics whereby miniaturization technologies often require the use of materials with high dielectric constants. The composition and chemical variations of materials such as solids and liquids can adopt characteristic responses, which are directly proportional to the amounts and types of chemical species added to the material. The examples given herein are related to aqueous suspensions whereby the electrical permittivity can be easily modulated via the addition of sodium chloride (NaCl).

Instrumentation

A common and reliable method for measuring the dielectric properties of liquid samples is to use an impedance analyzer in conjunction with a dielectric probe. The impedance analyzer directly measures the complex impedance of the sample under test and is then converted to permittivity using the system software. There are many methods used for measuring impedance, each of which has their own inherent advantages and disadvantages and factors associated with that particular method. Such factors include frequency range, measurement accuracy, and ease of operation. Common impedance measurements include bridge method, resonant method, current-voltage (I-V) method, network analysis method, auto-balancing bridge method, and radiofrequency (RF) I-V method. The RF I-V method used herein has several advantages over previously mentioned methods such as extended frequency coverage, better accuracy, and a wider measured impedance range. The principle of the RF I-V method is based on the linear relationship of the voltage-current ratio to impedance, as given by Ohm's law,

$$V = IZ$$

where V is voltage, I is current, and Z is impedance. This results in the impedance measurement sensitivity being constant regardless of measured impedance. Although a full description of this method involves circuit theory and is outside the scope of this module a brief schematic overview of the measurement principles is shown in Figure 14.3.

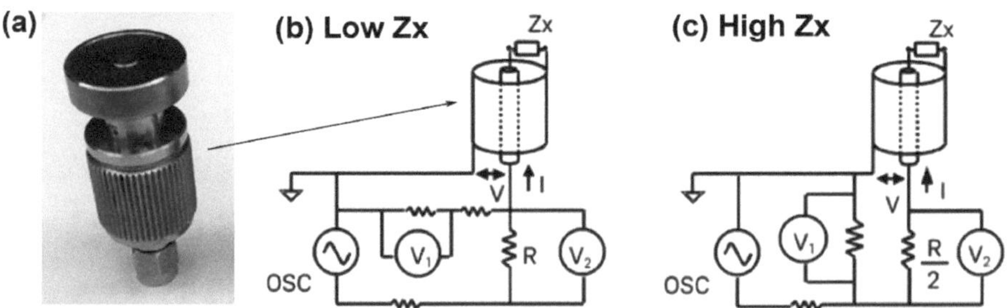

Figure 14.3: (a) Dielectric probe (liquids are placed on this probe). Circuit schematic of impedance measurements for (b) low and (c) high impedance materials. Circuit symbols Osc, Z$_x$, V, I, and R represent oscillator (i.e., frequency source), sample impedance, voltage, current, and resistance, respectively.

As can be seen in Figure 14.3, the RF I-V method, which incorporates the use of a dielectric probe, essentially measures variations in voltage and current when a sample is placed on the dielectric probe. For the low- impedance case, the impedance of the sample (Z_x) is given by,

$$Z_x = V/I = \frac{2R}{(V_2/V_1)-1}$$

for a high-impedance sample, the impedance of the sample (Z_x) is given by,

$$Z_x = V/I = (R/2)[(V_1/V_2)-1]$$

The instrumentation and methods described herein consist of an Agilent E4991A impedance analyzer connected to an Agilent 85070E dielectric probe kit. The impedance analyzer directly measures the complex impedance of the sample under test by measuring either the frequency-dependent voltage or current across the sample. These values are then converted to permittivity values using the system software.

Applications

Electrical permittivity of DI and saline water

In order to acquire the electrical permittivity of aqueous solutions the impedance analyzer and dielectric probe must first be calibrated. In the first instance, the impedance analyzer unit is calibrated under open- circuit, short-circuit, 50 Ω load, and low loss capacitance conditions by attaching the relevant probes shown in Figure 12.4. The dielectric probe is then attached to the system and re-calibrated in open-air, with an attached short circuit probe, and finally with 500 μL of highly purified deionized water (with a resistivity of 18.2 MΩ/cm at 25 °C) (Figure 14.5). The water is then removed, and the system is ready for acquiring data.

In order to maintain accurate calibration only the purest deionized water with a resistivity of 18.2 MΩ/cm at 25 °C should be used. To perform an analysis simply load the dielectric probe with 500 μL of the sample and click on the `acquire data' tab in the software. The system will perform a scan across the frequency range 200 MHz 3 GHz and acquire the real and imaginary parts of the complex permittivity.

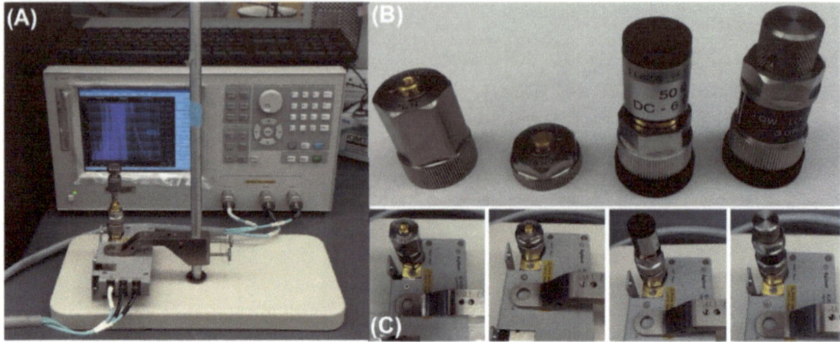

Figure 14.4: Impedance analyzer calibration (a) Agilent E4991A impedance analyzer connected to 85070E dielectric probe, (b) calibrations standards (left-to-right: open circuit, short circuit, 50 Ω load, low-loss capacitor), and (c) attachment of the open circuit, short circuit, 50 ohm load, and low-loss capacitor (left-to-right, respectively).

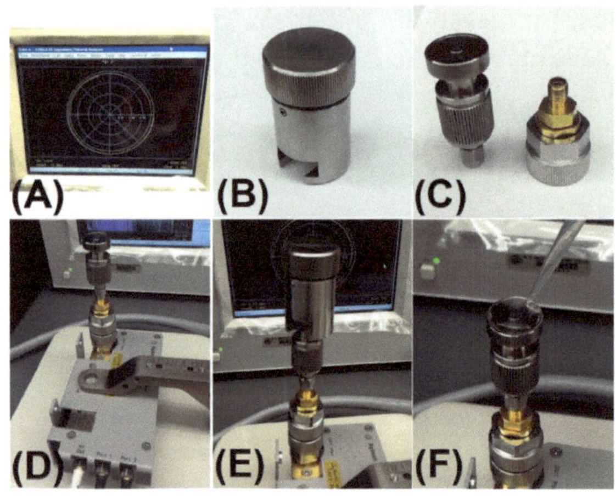

Figure 14.5: Dielectric probe calibration. (A) Impedance analyzer screen shot showing data line for dielectric probe in open-air. (B) short circuit probe (C) dielectric probe (D) dielectric probe connected to impedance analyzer under open air conditions (E) short-circuit probe attached to dielectric probe (F) 500 µL of deionized water on dielectric probe.

The period with which a data point is taken as well as the scale (i.e., log or linear) can also be altered in the software if necessary. To analyze another sample, remove the liquid and gently dry the dielectric probe with a paper towel. An open-air refresh calibration should then be performed (by pressing the relevant button in the software) as this prevents errors and instrument drift from sample to sample. To analyze a normal saline (0.9 % NaCl w/v) solution, dissolve 8.99 g of NaCl in 1 L of DI water (18.2 MΩ/cm at 25 °C) to create a 154 mM NaCl solution (equivalent to a 0.9 % NaCl w/v solution). Load 500 µL of the sample on the dielectric probe and acquire a new data set as mentioned previously.

Data analysis

The data les extracted from the impedance analyzer and dielectric probe setup previously described can be opened using any standard data processing software such as Microsoft Excel. The data will appear in three columns, which will be labeled frequency (Hz), ε', and ε'' (representing the real and imaginary components of the permittivity, respectively). Any graphing software can be used to create simple graphs of the complex permittivity versus frequency. In the example below (Figure 14.6) the real and complex permittivity's versus frequency (200 MHz 3 GHz) for the water and saline samples are plotted. For this frequency range no error correction is needed. For the analysis of frequencies below 200 MHz down to 10 MHz, which can be achieved using the impedance analyzer and dielectric probe configuration, error correction algorithms are needed to take into account electrode polarization effects that skew and distort the data.

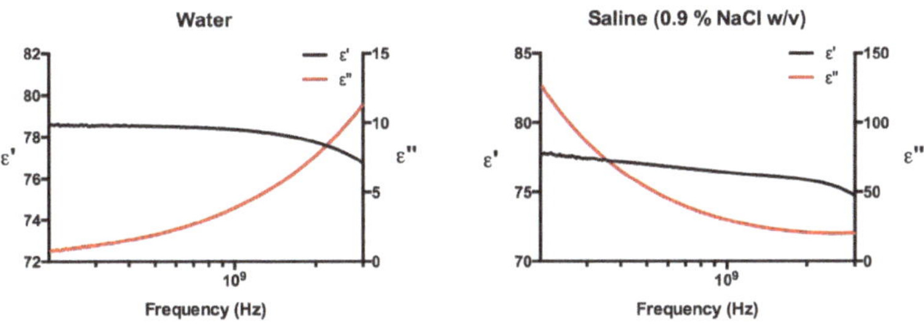

Figure 14.6: Real (black) and imaginary (red) components of permittivity for water (left) and saline (right) samples across the frequency range 200 MHz 3 GHz.

Bibliography

S. J. Corr, M. Raoof, B. T. Cisneros, A. W. Orbaek, M. A. Cheney, A. R. Barron, L. J. Wilson, and S. A. Curley, Radiofrequency electric-field heating behaviors of highly enriched semiconducting and metallic single-walled carbon nanotubes. *Nano Res.*, 2015, **8**, 2859.

A. C. Metaxas *Foundations of Electroheat: A Unified Approach*, Wiley Publishing (1996).

Measuring the Permittivity and Permeability of Lossy Materials: Solids, Liquids, Metals, Building Materials, and Negative-Index Materials. National Institute for Standards and Technology Tech. Note 1536 (2005).

Impedance Measurement Handbook. 6th edn., Keysight Technologies, Santa Rosa, CA (2020).

N. C. Lara, A. A. Haider, J. C. Ho, L. J. Wilson, A. R. Barron, S. A. Curley, and S. J. Corr, Water-structuring molecules and nanomaterials enhance radiofrequency heating in biologically relevant solutions. *Chem. Commun.*, 2016, **52**, 12630.

M. A. Laughton and D. F. Warne, *Electrical Engineers Reference Book*, Newnes Publishing, New York (2002).

H. Michael Gach and T. Nair, Radiofrequency interaction with conductive colloids: permittivity and electrical conductivity of single-wall carbon nanotubes in saline. *Bioelectromagnetics*, 2010, **31**, 582.

Chapter 15: Dynamic Mechanical Analysis

Julia Zhao and Andrew R. Barron

Introduction

Dynamic mechanical analysis (DMA), also known as forced oscillatory measurements and dynamic rheology, is a basic tool used to measure the viscoelastic properties of materials (particularly polymers). To do so, DMA instrument applies an oscillating force to a material and measures its response; from such experiments, the viscosity (the tendency to flow) and stiffness of the sample can be calculated. These viscoelastic properties can be related to temperature, time, or frequency. As a result, DMA can also provide information on the transitions of materials and characterize bulk properties that are important to material performance. DMA can be applied to determine the glass transition of polymers or the response of a material to application and removal of a load, as a few common examples. The usefulness of DMA comes from its ability to mimic operating conditions of the material, which allows researchers to predict how the material will perform.

A brief history

Oscillatory experiments have appeared in published literature since the early 1900s and began with rudimentary experimental setups to analyze the deformation of metals. In an initial study, the material in question was hung from a support, and torsional strain was applied using a turntable. Early instruments of the 1950s from manufacturers Weissenberg and Rheovibron exclusively measured torsional stress, where force is applied in a twisting motion.

Due to its usefulness in determining polymer molecular structure and stiffness, DMA became more popular in parallel with the increasing research on polymers. The method became integral in the analysis of polymer properties by 1961. In 1966, the revolutionary torsional braid analysis was developed; because this technique used a fine glass substrate imbued with the material of analysis, scientists were no longer limited to materials that could provide their own support. Using torsional braid analysis, the transition temperatures of polymers could be determined through temperature programming. Within two decades, commercial instruments became more accessible, and the technique became less specialized. In the early 1980s, one of the first DMAs using axial geometries (linear rather than torsional force) was introduced.

Since the 1980s, DMA has become much more user-friendly, faster, and less costly due to competition between vendors. Additionally, the developments in computer technology have allowed easier and more efficient data processing. Today, DMA is offered by most vendors, and the modern instrument is detailed in the Instrumentation section.

Basic principles of DMA

DMA is based on two important concepts of stress and strain. Stress (σ) provides a measure of force (F) applied to area (A),

$$\sigma = F/A$$

Stress to a material causes strain (γ), the deformation of the sample. Strain can be calculated by dividing the change in sample dimensions (ΔY) by the sample's original dimensions (Y),

$$\gamma = \Delta Y/Y$$

This value is often given as a percentage of strain.

The modulus (E), a measure of stiffness, can be calculated from the slope of the stress-strain plot, Figure 15.1, as displayed in,

$$E = \sigma/\gamma$$

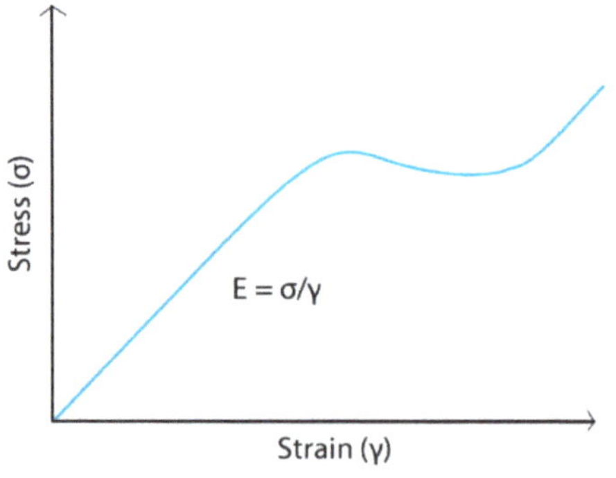

Figure 15.1: An example of a typical stress versus strain plot.

This modulus is dependent on temperature and applied stress. The change of this modulus as a function of a specified variable is key to DMA and determination of viscoelastic properties. Viscoelastic materials such as polymers display both elastic properties characteristic of solid materials and viscous properties characteristic of liquids; as a result, the viscoelastic properties are often a compromise between the two extremes. Ideal elastic properties can be related to Hooke's spring, while viscous behavior is often modeled using a dashpot, or a motion-resisting damper.

Creep-recovery

Creep-recovery testing is not a true dynamic analysis because the applied stress or strain is held constant; however, most modern DMA instruments have the ability to run this analysis. Creep-recovery tests the deformation of a material that occurs when load applied and removed. In the creep portion of this analysis, the material is placed under immediate, constant stress until the sample equilibrates. Recovery then measures the stress relaxation after the stress is removed. The stress and strain are measured as functions of time. From this method of analysis, equilibrium values for viscosity, modulus, and compliance (willingness of materials to deform; inverse of modulus) can be determined; however, such calculations are beyond the scope of this review.

Creep-recovery tests are useful in testing materials under anticipated operation conditions and long test times. As an example, multiple creep-recovery cycles can be applied to a sample to determine the behavior and change in properties of a material after several cycles of stress.

Dynamic testing

DMA instruments apply sinusoidally oscillating stress to samples and causes sinusoidal deformation. The relationship between the oscillating stress and strain becomes important in determining viscoelastic properties of the material. To begin, the stress applied can be described by a sine function where σo is the maximum stress applied, ω is the frequency of applied stress, and t is time. Stress and strain can be expressed with,

$$\sigma = \sigma_o \sin(\omega t + \delta)$$

$$\gamma = \gamma_o s \cos(\omega t)$$

The strain of a system undergoing sinusoidally oscillating stress is also sinusoidal, but the phase difference between strain and stress is entirely dependent on the balance between viscous and elastic properties of the material in question. For ideal elastic systems, the strain and stress are completely in phase, and the phase angle (δ) is equal to 0. For viscous systems, the applied stress leads the strain by 90°. The phase angle of viscoelastic materials is somewhere in between (Figure 15.2).

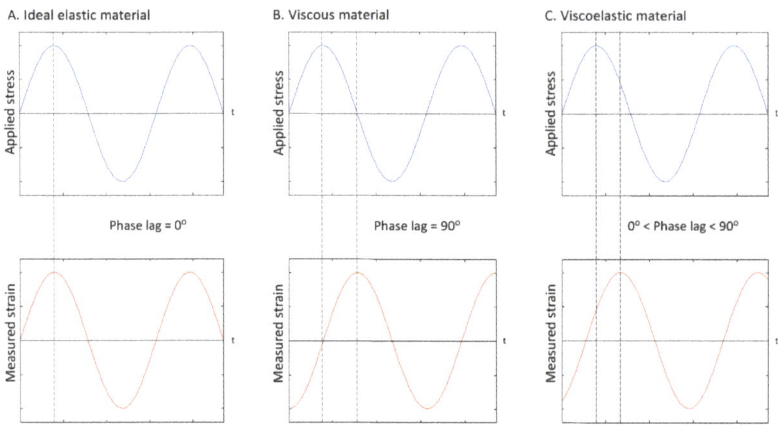

Figure 15.2: Applied sinusoidal stress versus time (above) aligned with measured stress versus time (below). (a) The applied stress and measured strain are in phase for an ideal elastic material. (b) The stress and strain are 90° out of phase for a purely viscous material. (c) Viscoelastic materials have a phase lag less than 90°. Image adapted from M. Sepe, *Dynamic Mechanical Analysis for Plastics Engineering*, Plastics Design Library: Norwich, NY (1998).

In essence, the phase angle between the stress and strain tells us a great deal about the viscoelasticity of the material. For one, a small phase angle indicates that the material is highly elastic; a large phase angle indicates the material is highly viscous. Furthermore, separating the properties of modulus, viscosity, compliance, or strain into two separate terms allows the analysis of the elasticity or the viscosity of a material. The elastic response of the material is analogous to storage of energy in a spring, while the viscosity of material can be thought of as the source of energy loss.

A few key viscoelastic terms can be calculated from dynamic analysis; their equations and significance are detailed in Table 15.1.

Term	Equation	Significance
Complex modulus (E*)	$E^* = E' + iE$	Overall modulus representing stiffness of material; combined elastic and viscous components
Elastic modulus (E')	$E' = (\sigma_o/\gamma_o)\cos\delta$	Storage modulus; measures stored energy and represents elastic portion
Viscous modulus (E)	$E = (\sigma_o/\gamma_o)\sin\delta$	Loss modulus; contribution of viscous component on polymer that flows under stress
Loss tangent (tanδ)	$\tan\delta = E/E'$	Damping or index of viscoelasticity; compares viscous and elastic moduli

Table 15.1: Key viscoelastic terms that can be calculated with DMA.

Types of dynamic experiments

A temperature sweep is the most common DMA test used on solid materials. In this experiment, the frequency and amplitude of oscillating stress is held constant while the temperature is increased. The temperature can be raised in a stepwise fashion, where the sample temperature is increased by larger intervals (e.g., 5 °C) and allowed to equilibrate before measurements are taken. Continuous heating routines can also be used (1 - 2 °C/minute). Typically, the results of temperature sweeps are displayed as storage and loss moduli as a function of temperature. For polymers, these results are indicative of polymer structure.

In time scans, the temperature of the sample is held constant, and properties are measured as functions of time, gas changes, or other parameters. This experiment is commonly used when studying curing of thermosets, materials that change chemically upon heating. Data is presented graphically using modulus as a function of time; curing profiles can be derived from this information. Frequency scans test a range of frequencies at a constant temperature to analyze the effect of change in frequency on temperature-driven changes in material. This type of experiment is typically run on fluids or polymer melts. The results of frequency scans are displayed as modulus and viscosity as functions of log frequency.

Instrumentation

The most common instrument for DMA is the forced resonance analyzer, which is ideal for measuring material response to temperature sweeps. The analyzer controls deformation, temperature, sample geometry, and sample environment.

Figure 15.3 displays the important components of the DMA, including the motor and driveshaft used to apply torsional stress as well as the linear variable differential transformer (LVDT) used to measure linear displacement. The carriage contains the sample and is typically enveloped by a furnace and heat sink.

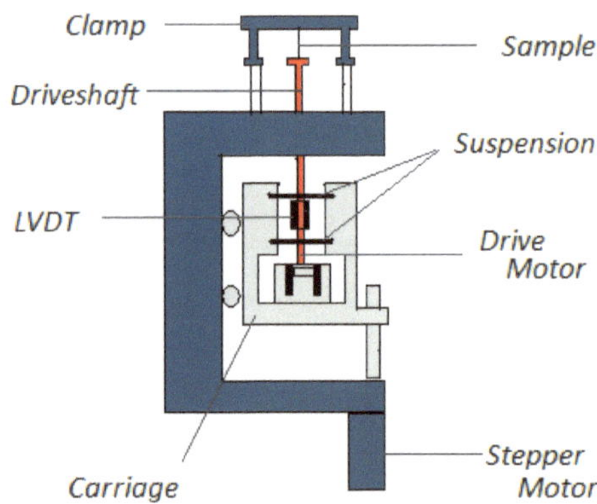

Figure 15.3: General schematic of DMA analyzer.

The DMA should be ideally selected to analyze the material at hand. The DMA can be either stress or strain controlled: strain-controlled analyzers move the probe a certain distance and measure the stress applied; strain-controlled analyzers provide a constant deformation of the sample (Figure 15.3). Although the two techniques are nearly equivalent when the stress-strain plot (Figure 15.1) is linear, stress-controlled analyzers provide more accurate results.

DMA analyzers can also apply stress or strain in two manners axial and torsional deformation (Figure 15.4). Axial deformation applies a linear force to the sample and is typically used for solid and semisolid materials to test flex, tensile strength, and compression. Torsional analyzers apply force in a twisting motion; this type of analysis is used for liquids and polymer melts but can also be applied to solids. Although both types of analyzers have wide analysis range and can be used for similar samples, the axial instrument should not be used for fluid samples with viscosities below 500 Pa-s, and torsional analyzers cannot handle materials with high modulus.

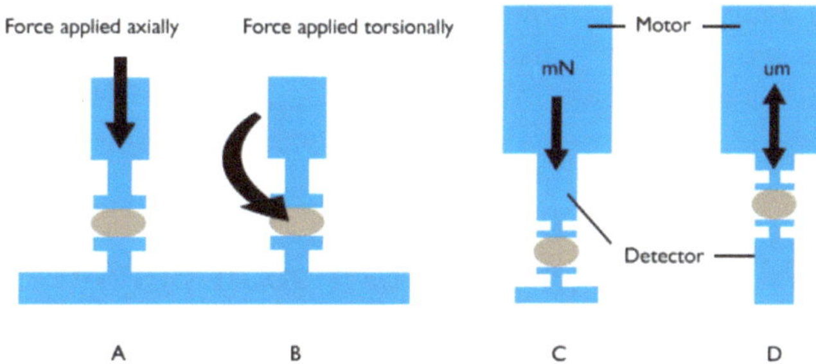

Figure 15.4: Types of DMA: (a) axially applied stress, (b) torsionally applied stress, (c) stress-controlled analyzer uses set movements, and (d) deformation is regulated in strain-controlled analyzers. Adapted from M. Sepe, *Dynamic Mechanical Analysis for Plastics Engineering*, Plastics Design Library: Norwich, NY (1998).

Different fixtures can be used to hold the samples in place and should be chosen according to the type of samples analyzed. The sample geometry affects both stress and strain and must be factored into the modulus calculations through a geometry factor. The fixture systems are specific to the type of stress application. Axial analyzers have a greater number of fixture options; one of the most commonly used fixtures is extension/tensile geometry used for thin films or fibers. In this method, the sample is held both vertically and lengthwise by top and bottom clamps, and stress is applied upwards (Figure 15.5).

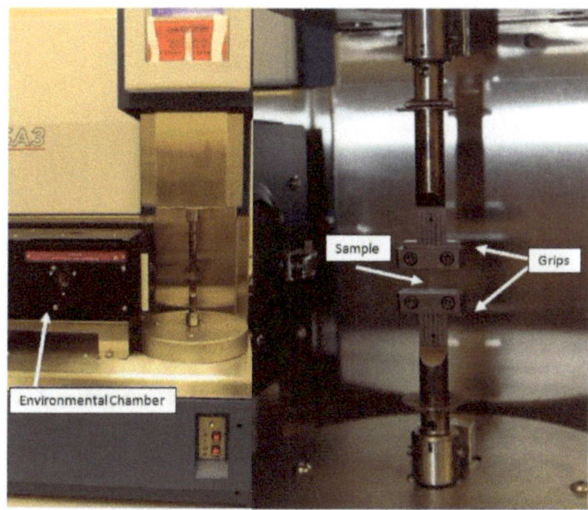

Figure 15.5: Axial analyzer with DMA instrument (left) and axial analyzer with extension/tensile geometry (right).

For torsional analyzers, the simplest geometry is the use of parallel plates. The plates are separated by a distance determined by the viscosity of the sample. Because the movement of the sample depends on its radius from the center of the plate, the stress applied is uneven; the measured strain is an average value.

DMA of the glass transition of polymers

As the temperature of a polymer increases, the material goes through a number of minor transitions ($T\gamma$ and T_β) due to expansion; at these transitions, the modulus also undergoes changes. The glass transition of polymers (T_g) occurs with the abrupt change of physical properties within 140-160 °C; at some temperature within this range, the storage (elastic) modulus of the polymer drops dramatically. As the temperature rises above the glass transition point, the material loses its structure and becomes rubbery before finally melting. The idealized modulus transition is pictured in Figure 15.6.

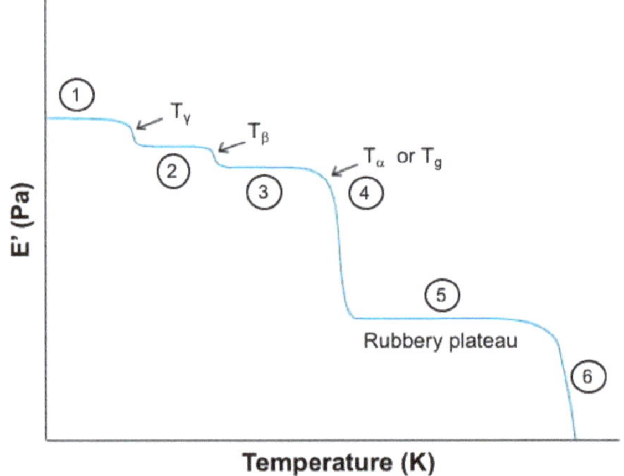

Figure 15.6: Ideal storage modulus transitions of viscoelastic polymers. (1) Local motion, (2) bend and stretch, (3) side groups, (4) gradual main chain, (5) large scale chain, and (6) chain slippage. Adapted from K. P. Menard, *Dynamic Mechanical Analysis: A Practical Introduction*, 2nd edn., CRC Press: Boca Raton, FL (2008).

The glass transition temperature can be determined using either the storage modulus, complex modulus, or tan δ (versus temperature) depending on context and instrument; because these methods result in such a range of values

(Figure 15.7), the method of calculation should be noted. When using the storage modulus, the temperature at which E' begins to decline is used as the T_g. Tan δ and loss modulus E show peaks at the glass transition; either onset or peak values can be used in determining T_g These different methods of measurement are depicted graphically in Figure 15.7.

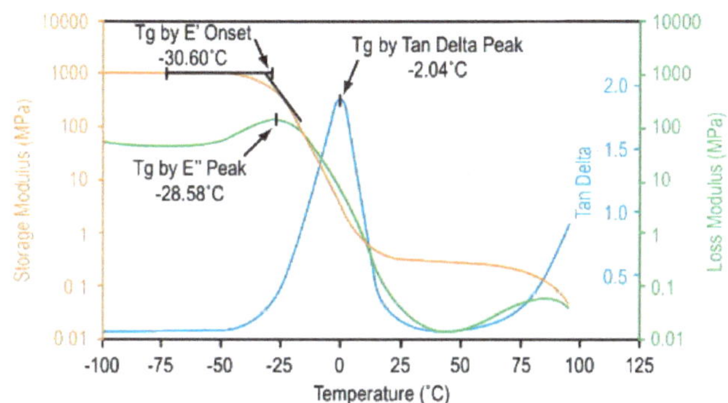

Figure 15.7: Different industrial methods of calculating glass transition temperature (T_g). Copyright 2014, TA Instruments. Used with permission.

Advantages and limitations of DMA

Dynamic mechanical analysis is an essential analytical technique for determining the viscoelastic properties of polymers. Unlike many comparable methods, DMA can provide information on major and minor transitions of materials; it is also more sensitive to changes after the glass transition temperature of polymers. Due to its use of oscillating stress, this method is able to quickly scan and calculate the modulus for a range of temperatures. As a result, it is the only technique that can determine the basic structure of a polymer system while providing data on the modulus as a function of temperature. Finally, the environment of DMA tests can be controlled to mimic real-world operating conditions, so this analytical method is able to accurately predict the performance of materials in use.

DMA does possess limitations that lead to calculation inaccuracies. The modulus value is very dependent on sample dimensions, which means large inaccuracies are introduced if dimensional measurements of samples are slightly inaccurate. Additionally, overcoming the inertia of the instrument used to apply oscillating stress converts mechanical energy to heat and changes the temperature of the sample. Since maintaining exact temperatures is important in temperature scans, this also introduces inaccuracies. Because data processing of DMA is largely automated, the final source of measurement uncertainty comes from computer error.

Bibliography

K. P Menard, *Dynamic Mechanical Analysis: A Practical Introduction*, 2nd edn., CRC Press: Boca Raton, FL (2008).

M. A. Meyer and K. K. Chawla, *Mechanical Properties of Materials*, 2nd edn., Cambridge University Press: New York (2009).

E. I. Rivin, *Stiffness and Damping in Mechanical Design*, Marcel Dekker: New York (1999).

G. Rotter and H. Ishida, Dynamic mechanical analysis of the glass transition: curve resolving applied to polymers. *Macromolecules*, 1992, **25**, 2170.

M. Sepe, *Dynamic Mechanical Analysis for Plastics Engineering*, Plastics Design Library: Norwich, NY (1998).

R. J. Young, *Introduction to Polymers*, 3rd edn., CRC Press: Boca Raton, FL (2011).